FASCINATING SCIENCE PROJECTS

ANIMAL LIFE

Sally Hewitt

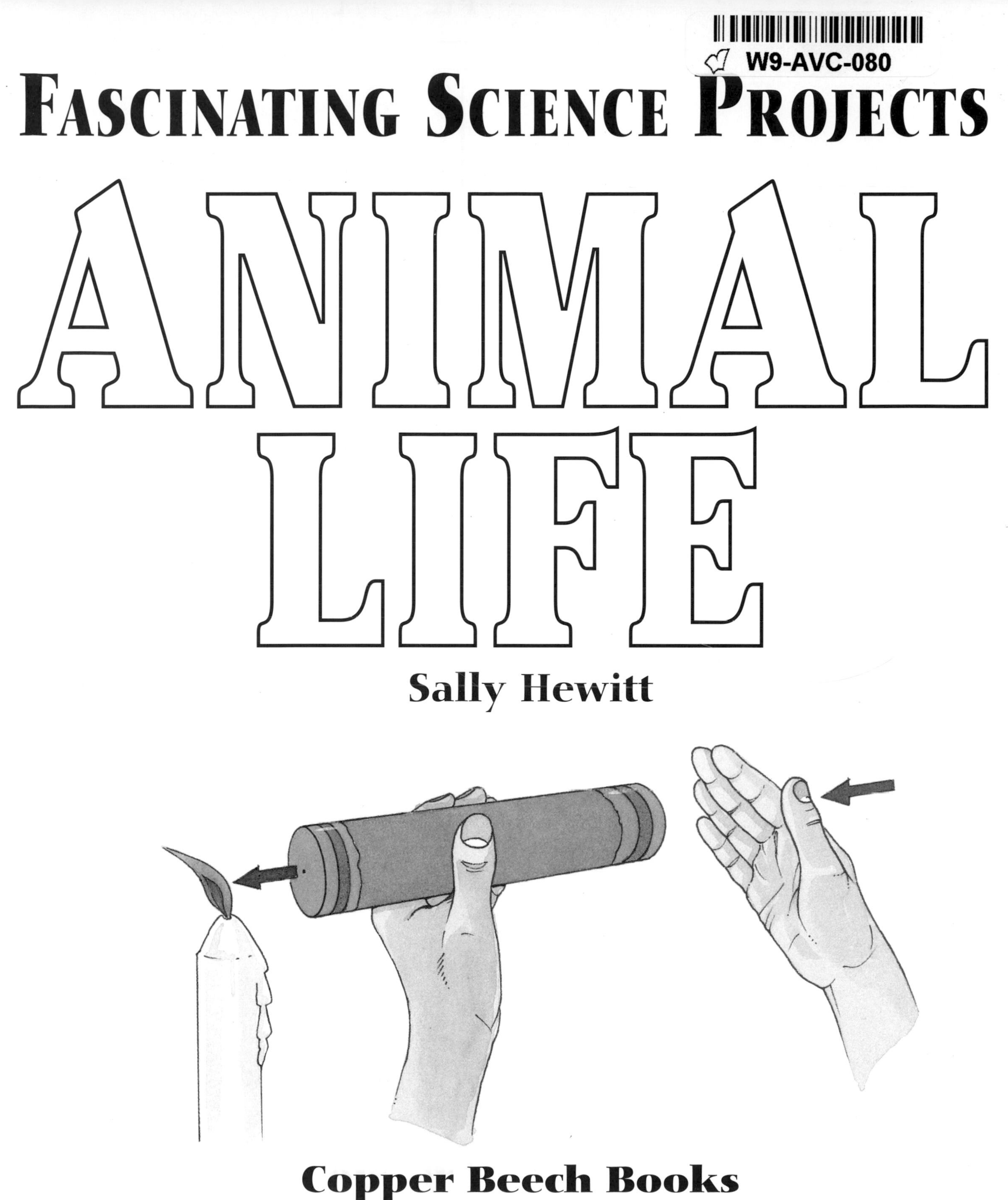

Copper Beech Books
Brookfield, Connecticut

Produced by
Aladdin Books Ltd
28 Percy Street
London W1P 0LD

ISBN 0-7613-2716-9 (lib. bdg.)
ISBN 0-7613-1631-0 (pbk.)

First published in the United States in 2002 by
Copper Beech Books,
an imprint of
The Millbrook Press
2 Old New Milford Road
Brookfield, Connecticut 06804

Designers:
Flick, Book Design and Graphics
Pete Bennett

Editor:
Harriet Brown

Illustrators:
Andrew Geeson,
Catherine Ward and Peter Wilks—SGA
Cartoons: Tony Kenyon—BL Kearley

Consultant:
Bryson Gore

Printed in Belgium

Cataloging-in-Publication data is on file
at the Library of Congress.

Contents

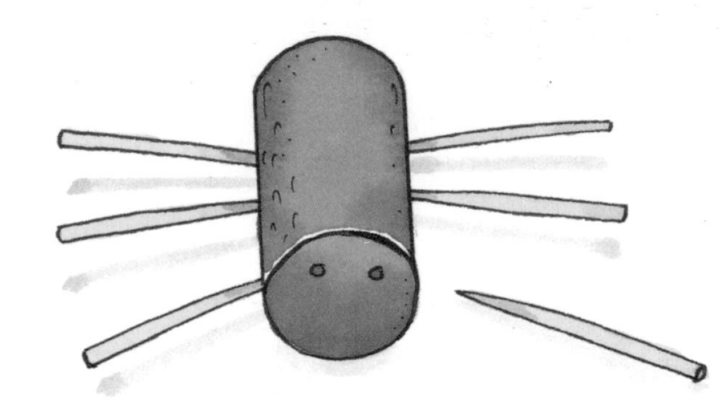

Introduction

In this book, animal life is explained through a series of fascinating projects and experiments. Each chapter deals with a different topic on animals, such as vertebrates or breathing, and contains a major project and other simple experiments, “Magic panels,” and “Fascinating fact” boxes. At the end of every chapter is a summary, explaining what has been shown and what it tell us. Projects requiring the use of sharp tools should be done under adult supervision.

This states the purpose of the project

METHOD NOTES
Helpful hints on things to remember when carrying out your project.

Materials
In this box is a full list of the items needed to carry out each main project.

1. The steps that describe how to carry out each project are listed clearly as numbered points.
2. Where there are illustrations to help you understand the instructions, the text refers to them as Figure 1, etc.

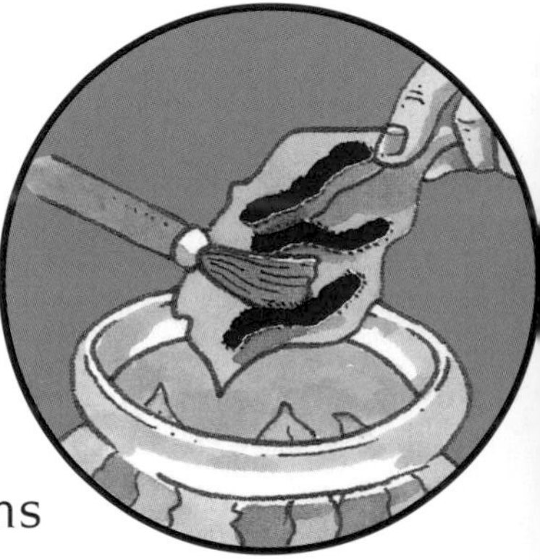

Figure 1

Figure 2

The Amazing Magic Panel

This heading explains what is happening

These boxes contain an activity or experiment that has a particularly dramatic or surprising result!

WHY IT WORKS
You can find out exactly what happened here, too.

WHAT THIS SHOWS

These boxes, which are headed either "What this shows" or "Why it works," contain an explanation of what happened during your project, why it happened, and the meaning of the result.

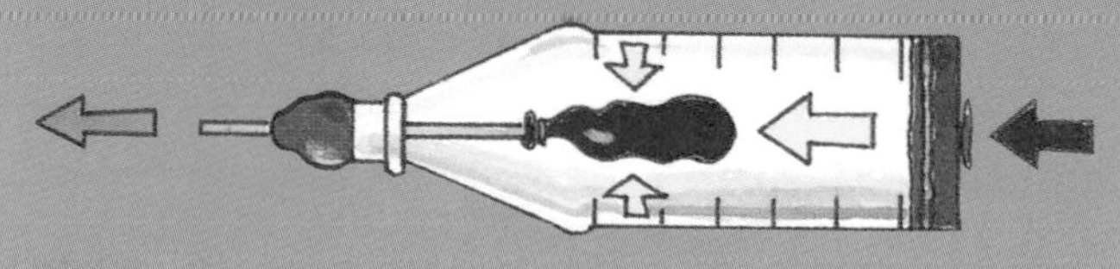

Fascinating facts!
An amusing or surprising fact related to the theme of the chapter.

Where the project involves using a sharp knife or anything else that requires adult supervision, you will see this warning symbol.

The text in these circles links the theme of the topic from one page to the next in the chapter.

What is animal life?

Everything around you is either a nonliving or a living thing. This book, your clothes, and the coins in your pocket are all nonliving things. Plants and animals, including human beings like you, are all living things. The animal kingdom is made up of millions of very different kinds of animals. Animals can be divided into two groups—those with a backbone, called vertebrates, and those without a backbone, called invertebrates.

Watch wriggling worms living in soil

Materials

- waterproof glue
- soil, grass, and leaves
- red cellophane
- a cloth
- a flashlight
- worms
- 3 pieces of wood 12 in (30 cm) by 1 in (2.5 cm)
- 2 pieces of clear hard plastic 14 in (35 cm) by 12 in (30 cm)

METHOD NOTES

Worms work in the dark, so cover them up and let them get busy.

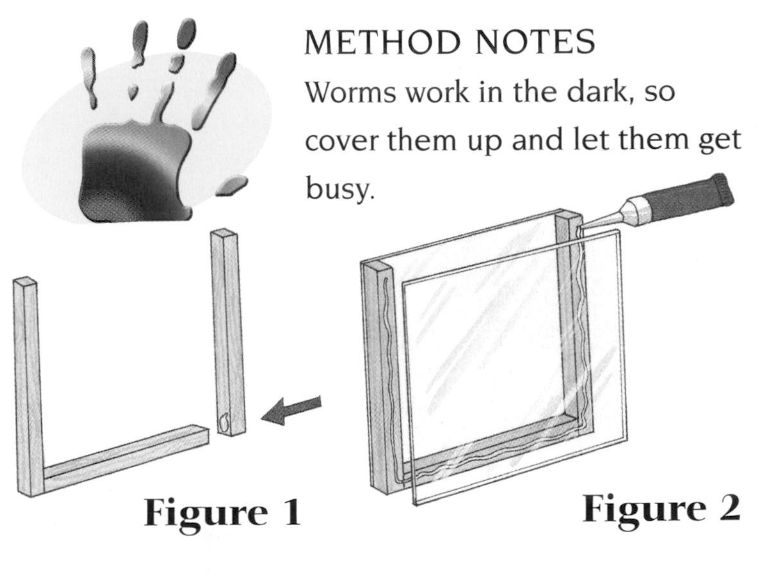

Figure 1 Figure 2

1. Glue the pieces of wood together with the waterproof glue to make a three-sided frame (Figure 1) for your worm habitat. Let the glue dry thoroughly.
2. Glue a sheet of plastic to each side of the frame (Figure 2).

Figure 3

3. Fill your habitat with soil. Make sure you keep the soil damp.
4. Obtain four or five big worms and put them on top of the soil (Figure 3).
5. Add some fresh grass and leaves. Cover the habitat opening with a cloth.
6. Cover a flashlight with red cellophane and watch the worms at work in the dark (Figure 4).

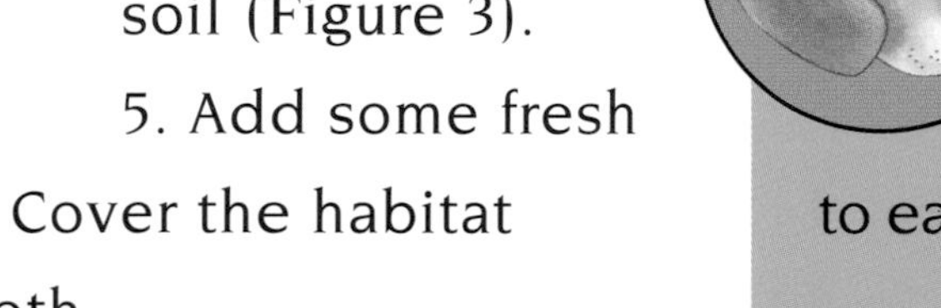

Figure 4

WHAT THIS SHOWS

As you observe worms living in soil, you will see some of the seven characteristics of animals, such as eating, excretion, and movement.

1. Breathing—Animals breathe in oxygen and breathe out carbon dioxide.
2. Eating—Animals need to eat to give them the energy to live and grow.
3. Excretion—Animals get rid of waste from eating.
4. Movement—Animals move around. Movement also goes on inside their bodies.
5. Senses—Animals have senses, for example, sight, hearing, smell, taste, and touch.
6. Growth—Animals grow throughout their lives.
7. Reproduction—Animals can make new living things like themselves.

What is animal life?

Zoologists are people who study animals. They have divided the animal kingdom into groups. You belong to a group of animals called mammals. All mammals have a backbone.

I'M ALIVE!

Make a checklist like the one on the right. Now make some toast, eat it and check if you are a living thing. Are you breathing and moving? Can you smell and taste the toast? Do you go to the toilet some time later?

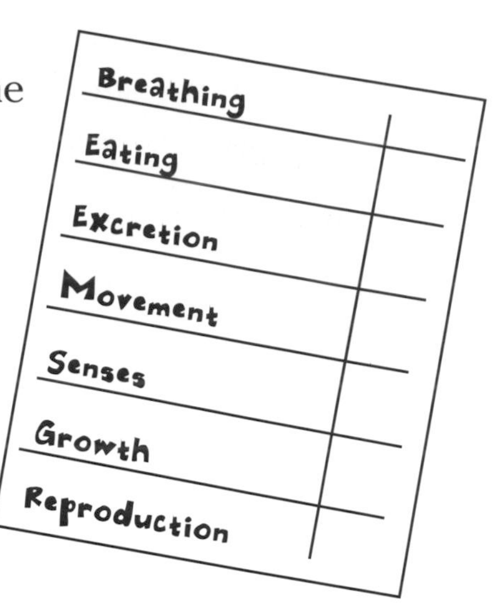

Figure 1

Food gives you energy to grow. When you are fully grown, you will be ready to reproduce. Check the boxes.

BACKBONES

The backbone of a fish is knobby like yours. Feel your friend's backbone. Can you count all 33 of the knobby backbones? These are called vertebrae.

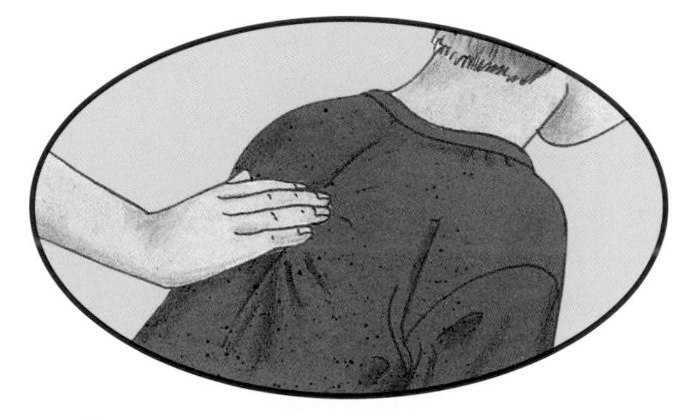

Too small to see!

Millions of animals on Earth are invertebrates that are too small to see. Everything around you is teeming with tiny creatures.

ANIMAL OR VEGETABLE

Make a collection of things like the ones shown below. Divide a large piece of paper into two sections with the headings "Living" and "Nonliving." Then divide the Living section into "Plants" and "Animals," and the Nonliving section into "Once alive" and "Never alive." Now sort each thing in your collection into its place on the chart.

Living		Nonliving	
Plants	Animals	Once alive	Never alive
pears	cat	leather shoes	stones
leaves	me	wooden pencil	jar
carrots		jelly	
lemon		paper	
plant			

WHAT THIS SHOWS

Everything around you is either a living or a nonliving thing. Animals have all the seven characteristics listed on page 7. A cat is an animal and a living thing. Fruit is part of a plant and a living thing. A stone is a nonliving thing that was never alive. Jelly is a nonliving thing that was once alive, because it is made from fruit that has been picked.

All animals, including humans like you, are living things that breathe, feed, get rid of waste, move, feel, grow, and can make new life like themselves.

Arthropods

Arthropods are a group of animals that don't have a backbone. There are more kinds of arthropods than all other kinds of animals put together. Your skeleton is inside your body. Arthropods have a "skeleton" outside their soft bodies, and it is like a coat of armor. This exoskeleton doesn't grow, so as arthropods get bigger, they molt their old coat and form a new one to fit them. Centipedes, lobsters, crabs, spiders, and insects are all arthropods.

Discover how some insects walk on water

Materials

- a clean bowl
- clean water
- a cork
- 3 toothpicks
- soap
- a pen

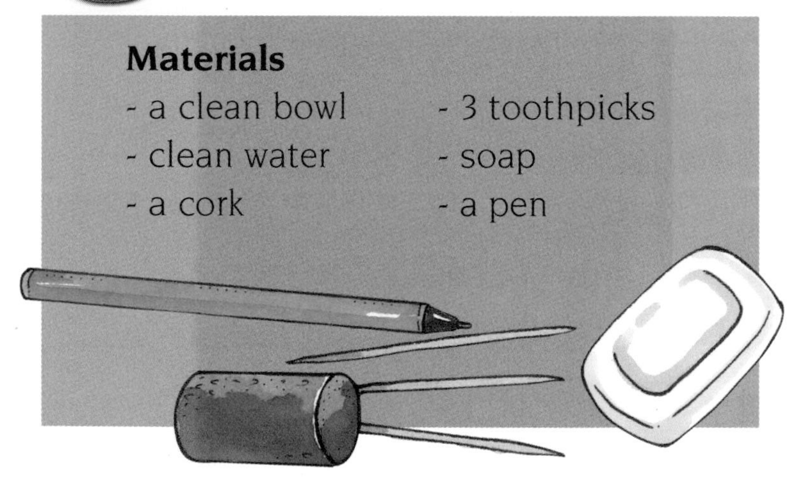

METHOD NOTES
Use a clean bowl and water with no trace of soap or detergent.

1. Fill the bowl with cold water from the faucet. Let the water settle and become quite still.

2. Break the toothpicks in half. Push 3 half-sticks into each side of the cork (Figure 1). Test to make sure that your "insect" balances. Draw on two black eyes.

Figure 1

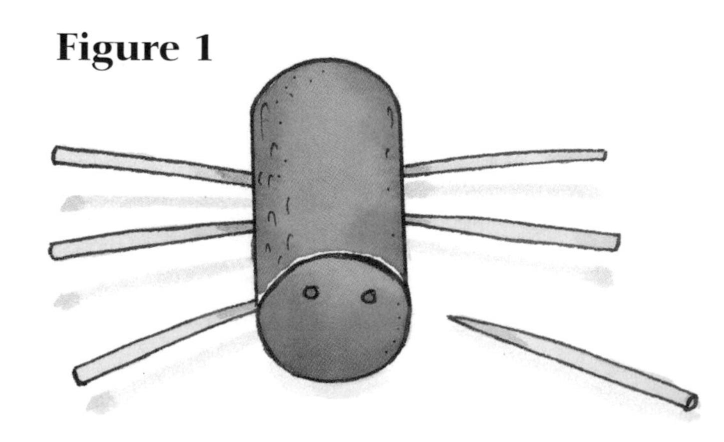

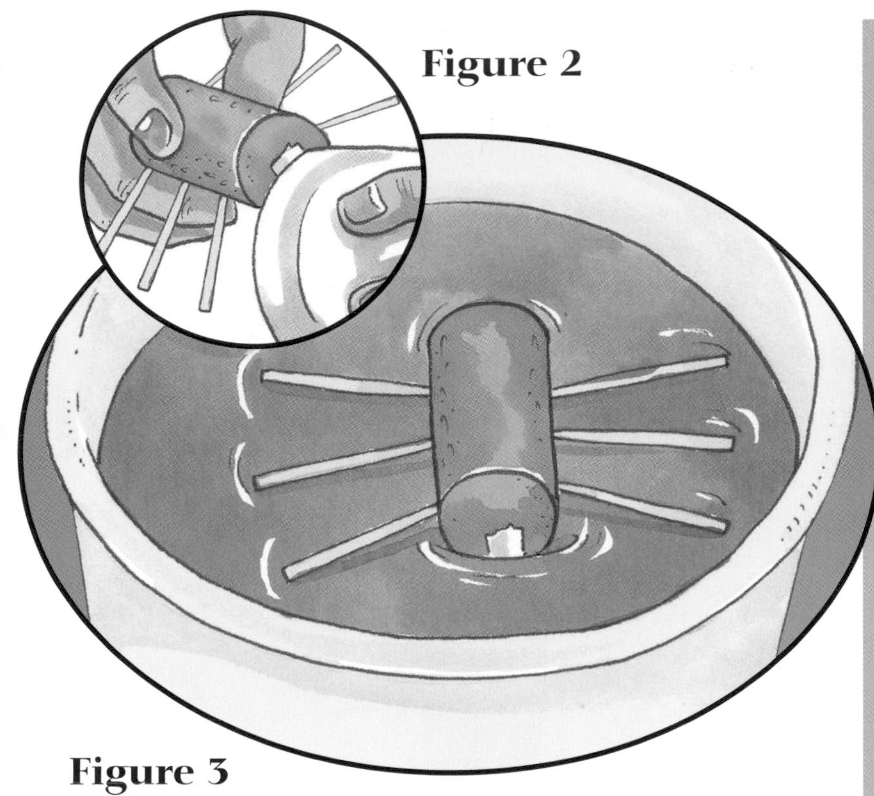

Figure 3

3. Rub a small amount of soap on the bottom end of the cork (Figure 2).
4. Gently place the "insect" in the water at the edge of the bowl (Figure 3) and watch it move across the water.

WHY IT WORKS

Water has a kind of skin formed by a force called surface tension. Normally, water insects skim across the surface of ponds without breaking this skin. But a water cricket in danger gives out a tiny bit of oil that breaks the surface tension behind it. The unbroken surface tension of the water in front pulls the water cricket forward and gets it out of trouble fast.

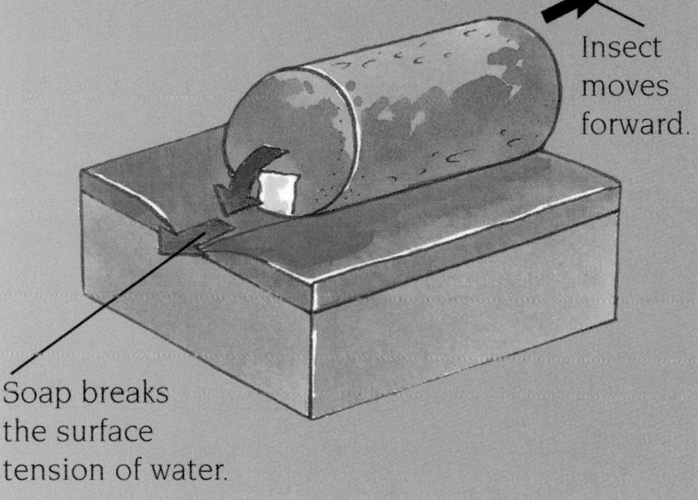

Amazing mayflies

A mayfly starts its life as a brown nymph living under water. After it breaks out of its skin and flies away on shiny wings, it lives only for a few hours—just long enough to mate and lay eggs.

Arthropods

You can tell an insect from other arthropods by counting its legs. All adult insects have six legs. Many young insects look very different from the adults they become.

PITFALL TRAP

Pitfall traps catch small creatures that slither and crawl along the ground. Dig a hole in some soil and insert a clean jar. Put fruit, jelly, or meat in it and cover the top with a stone, leaving a small gap (Figure 1). Leave it overnight. In the morning, identify the creatures you have caught before you let them go.

Figure 1

On a warm night, shine a flashlight outside and see which creatures are attracted to the light.

THE AMAZING SILK TRAP

See how spiders use their webs to catch prey

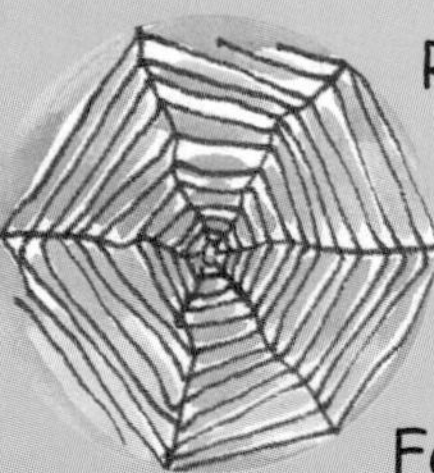

Run a thread across a plate and put a tempting doughnut on top. Pull the thread tight, hold onto it, and hide away. Feel the vibrations in the thread when someone or something touches it.

WHAT THIS SHOWS

Vibrations travel along the thread to tell you that something is there. Spiders weave intricate silk webs to catch insects for food. The spider hides under a leaf with one foot on a line attached to its web. Vibrations caused by the insect alert the spider, and it rushes out to catch its prey.

KEEPING CATERPILLARS

Look for caterpillars and gently brush them into a box. Put some of the leaves they were feeding on in the box. Take them home. Clean a large jar and carefully punch holes in the lid with a skewer to let in air. Cover the bottom with soil and add a stone and a small jar of water for the leaves. Put the caterpillars into the jar.

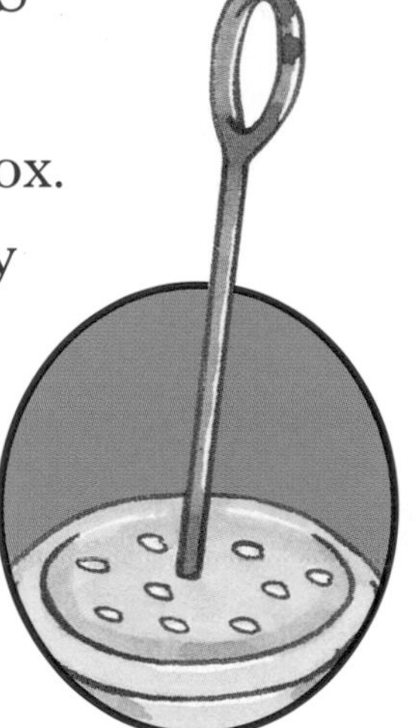

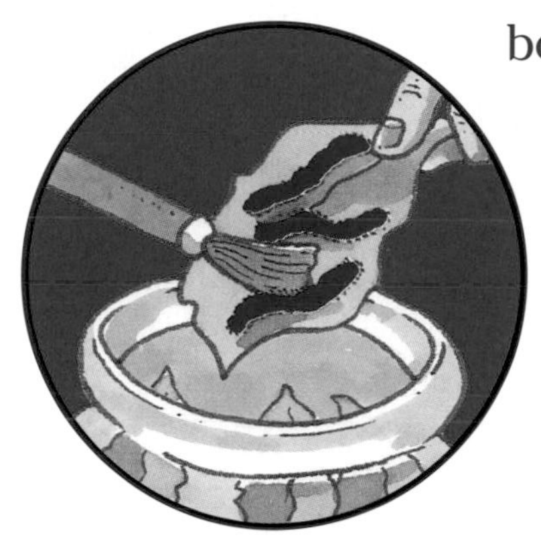

WHAT THIS SHOWS

The caterpillars in your jar feed on leaves (1). Each caterpillar can become a chrysalis (2). Inside the chrysalis, the caterpillar changes shape (3). This is called metamorphosis. The chrysalis opens, revealing a butterfly (4)! A male butterfly then mates with a female, who lays her eggs on a leaf (5).

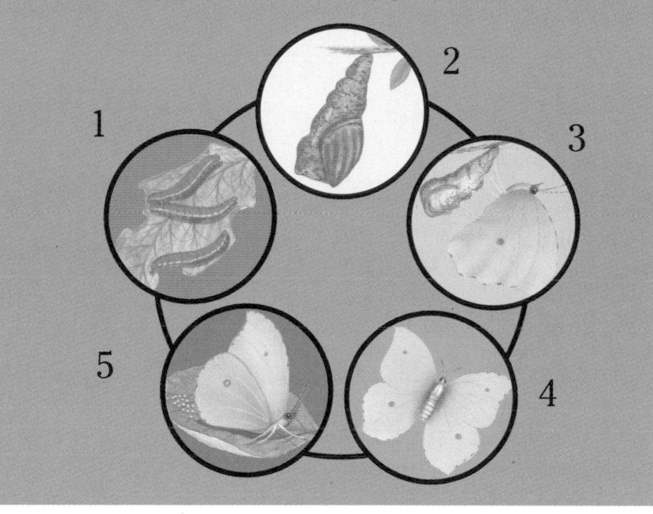

Give them a fresh supply of leaves every day for food. Keep them in a cool, shady place where you can watch them feed and grow. They may eventually form a chrysalis, from which a butterfly emerges.

Arthropods are a huge group of animals without backbones. They have specialized ways of getting food and escaping from danger. Some arthropods dramatically change the way they look as they develop.

Vertebrates

Vertebrates are animals with a backbone. They are divided into five groups—fish, amphibians, reptiles, birds, and mammals. Fish are covered in scales and spend their lives under water. Amphibians spend some of their lives on land and some under water, and most have a smooth skin. Some reptiles live on land and some live in water. They have scaly skin. Birds have feathers, and mammals have hair or fur. Most mammals give birth to live young, but all other vertebrates lay eggs.

Take your temperature

METHOD NOTES

Find 98.6°F (37°C), your normal temperature, on the thermometers.

Materials
- a pencil
- paper
- a strip thermometer
- a digital thermometer

1. Make sure the strip thermometer is all one color before you take each temperature (Figure 1). Make sure the digital thermometer is reset before you begin.
2. Draw a picture of yourself showing your head, arms, legs, fingers, and toes (Figure 2).

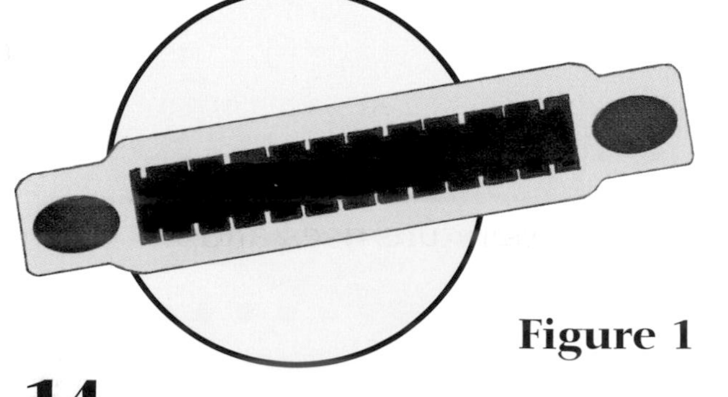

Figure 1

Figure 2

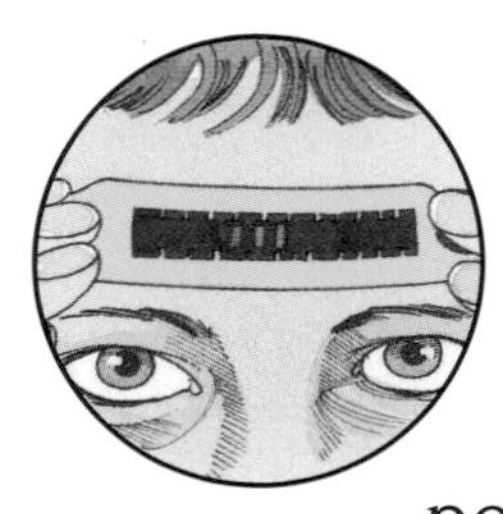

3. When you are feeling warm, take your temperature at different points on your body. Use the strip thermometer for your forehead and your lips. Use the digital one for under your arms, behind your knees, your mouth, your fingertips, and between your toes.

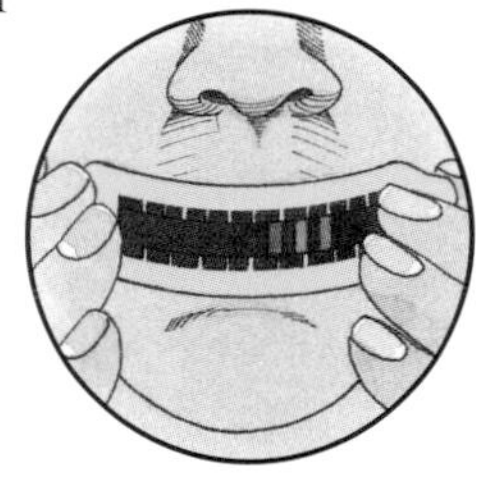

4. Take your temperature at night before you go to bed and in the morning as soon as you wake up.

5. Record your temperature readings on the picture of yourself.

WHAT THIS SHOWS

The internal temperature of humans and mammals stays about the same all the time. The temperature in your mouth is normally 98.6°F (37°C). Your fingertips and toes can feel cold or hot. They cool down or heat up to save or give out heat, which keeps the inside of your body at the same temperature.

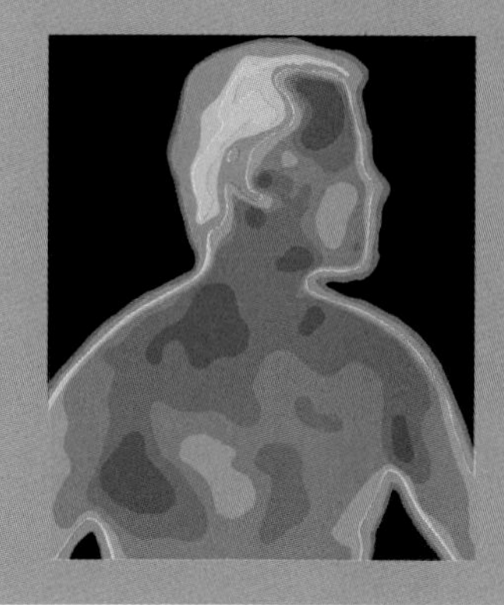

Do you notice any differences between the temperature on different parts of your body and at different times of day?

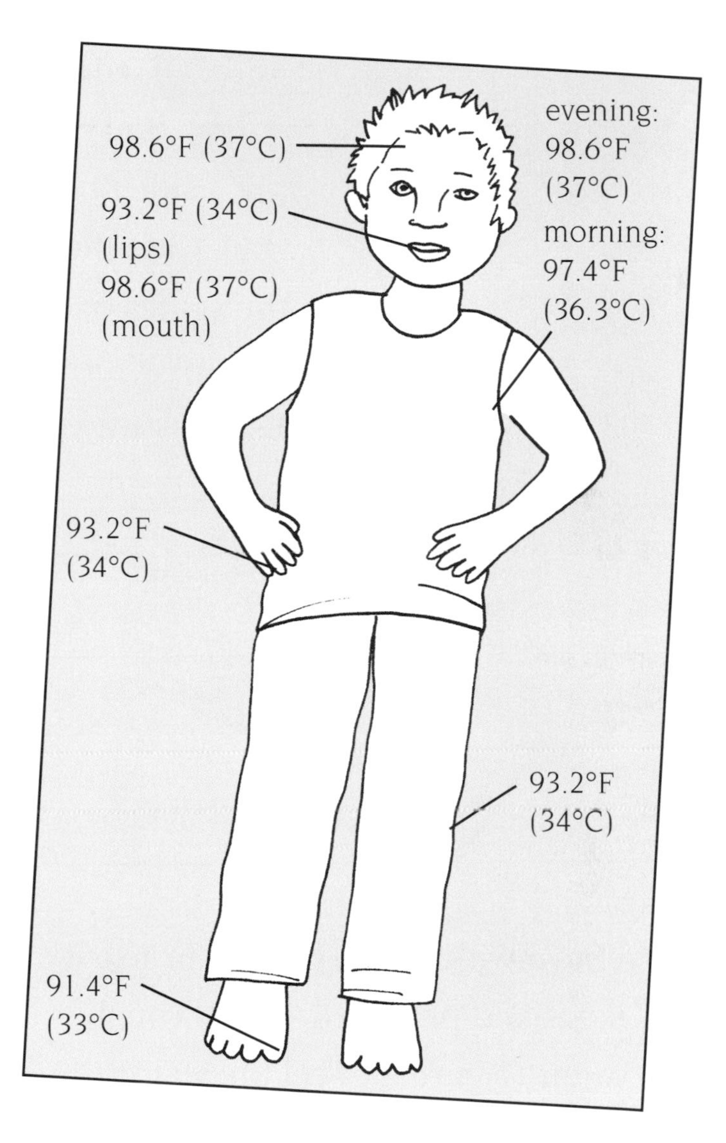

6. Take your temperature on cold days and on hot days, and see if your temperature changes. If you have a fever, your temperature rises and you feel burning hot.

Vertebrates

Vertebrates have many different kinds of coverings. Fish and reptiles have scales, birds have feathers, and most mammals have hair or fur.

OVERLAPPING FISH SCALES

Cut a fish shape and lots of scales out of shiny cardboard (Figure 1). Put waterproof glue on one edge of each scale. Glue a row of scales at the fish's tail end, with the unglued side pointing toward the tail. Glue the next row above it so the scales overlap (Figure 2). Cover the whole body of the fish this way.

Figure 1

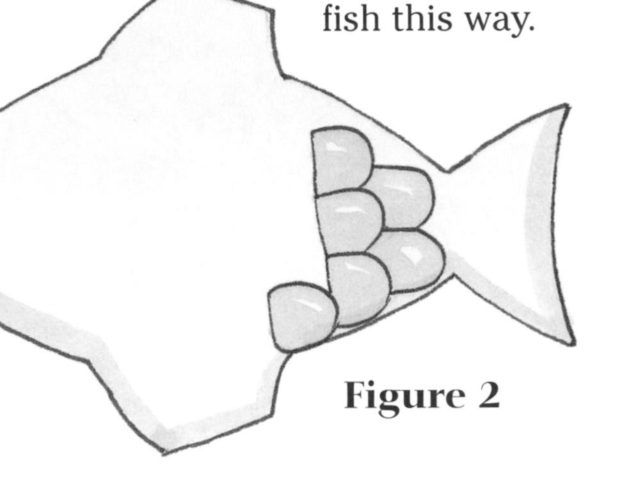

Figure 2

WHAT THIS SHOWS

Fish are designed for swimming. They have very streamlined bodies. Their scales are arranged in an overlapping pattern so that water slips over them easily.

Fill a bowl with water. Hold your fish by the nose, and pull it forward through the water (Figure 3). Feel the water slip easily over the scales. Now hold the fish by the tail and pull it backward. The water will catch in the scales and make it difficult to pull.

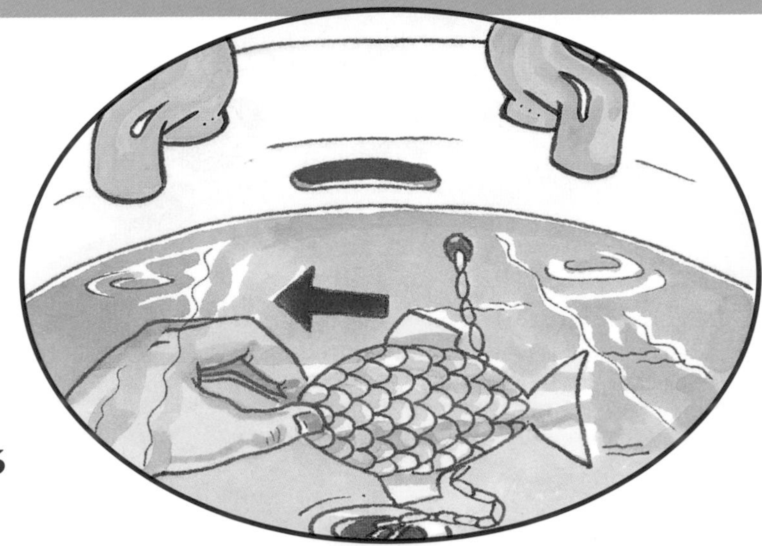

Figure 3

The Amazing Strong Hair

Test the strength of your hair

Find someone with long hair, and ask an adult to cut one or two hairs. Stick one end of a hair to a small plastic bag with tape and tie the other end around a pencil. Let the bag hang in the air and drop in marbles one by one. It's amazing how many marbles one hair will hold before it breaks.

WHY IT WORKS
Hair is surprisingly strong. It is made from the same material as fingernails. Mammals need strong, tough hair for protection from the weather. Compare how many marbles one hair can hold with how many a thread will hold.

WATERPROOF FEATHERS

Run your finger and thumb up and down the shaft of a feather to lock and unlock its barbs. Dip the feather into a bowl of water mixed with food coloring. Shake the feather to see how it repels the colored water. Birds use their beaks to smooth oil from glands onto their feathers. This keeps the feathers waterproof and in good condition for flying.

Try dipping a paper towel and a plastic bag into the bowl to see which materials absorb the coloring and which are waterproof.

Vertebrates are a large group of animals with backbones. Their different skin coverings help them survive. Feathers help birds to fly, scales help fish to be streamlined, and fur and hair keep mammals warm and dry.

Breathing

All living things need air to breathe. Mammals breathe air in and out of their lungs. They take in oxygen from the air and breathe out carbon dioxide as waste. Sea mammals, such as whales and dolphins, can hold their breath under water for a very long time, but they eventually have to come up to the surface for air. Most fish don't have lungs. They take in oxygen from the water through their gills. Insects breathe through tiny holes in their exoskeletons called spiracles.

Make a model lung and diaphragm to see how you breathe

METHOD NOTES
Use a strong bottle that won't collapse when you cut it.

Materials
- a clear plastic bottle
- a small balloon
- a large balloon
- modeling clay
- a drinking straw
- a small rubber band
- a large rubber band
- scissors

Figure 1

1. Ask an adult to help you cut the bottom off a plastic bottle.
2. Knot the end of the large balloon and cut off the top. Stretch the opening over the cut end of the bottle and keep it in place with a large rubber band (Figure 1).

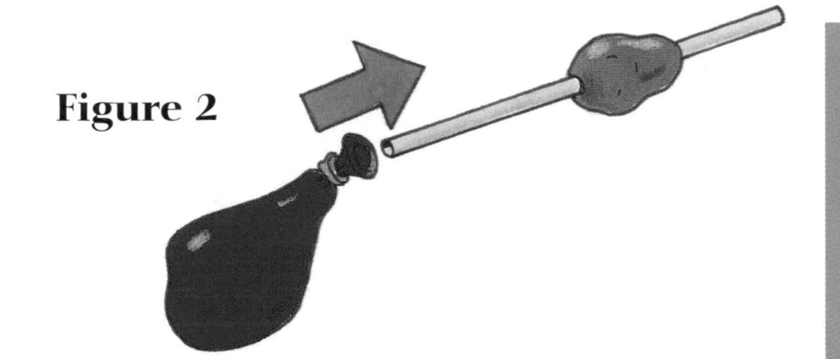

Figure 2

3. Attach a small balloon to the end of the straw with a small rubber band (Figure 2).
4. Put the straw into the bottle and use the modeling clay to fit it tightly into the bottle neck (Figure 3).

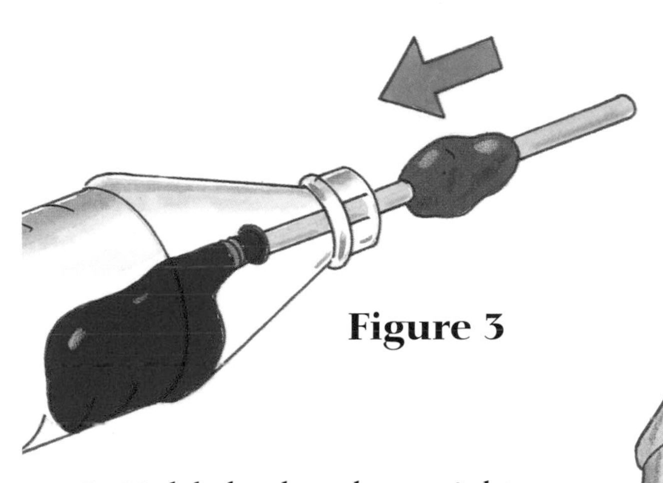

Figure 3

5. Hold the bottle upright with one hand. Use the other hand to pull down on the knot of the balloon stretched over the bottle (Figure 4). Watch the small balloon fill with air.
6. Push the knot up and watch the small balloon deflate.

Figure 4

WHAT THIS SHOWS

Imagine that the opening of the straw is your nose and the small balloon is one of your lungs. The balloon stretched over the bottle is your diaphragm—a large muscle at the bottom of your chest. The bottle is your chest. When you breathe in, your diaphragm relaxes and draws air into your lungs, which stretch as they fill up. When you breathe out, it tightens and pushes air out of your lungs, and they go back to their normal size. Breathe out and measure your chest. Take a deep breath and measure it again. By how much has it expanded (got bigger)?

Breathing

You breathe in air that is rich in oxygen. Oxygen travels from your lungs to all the different parts of your body that need it. You breathe out moisture and a waste gas called carbon dioxide.

AIR BREATHED IN AND OUT

A candle needs oxygen to burn. Without oxygen, the flame will go out. Light a candle and cover it with a glass jar (Figure 1). Time how long the candle burns before it uses up all the oxygen in the jar and goes out. You must wear oven gloves to pick up the jar.

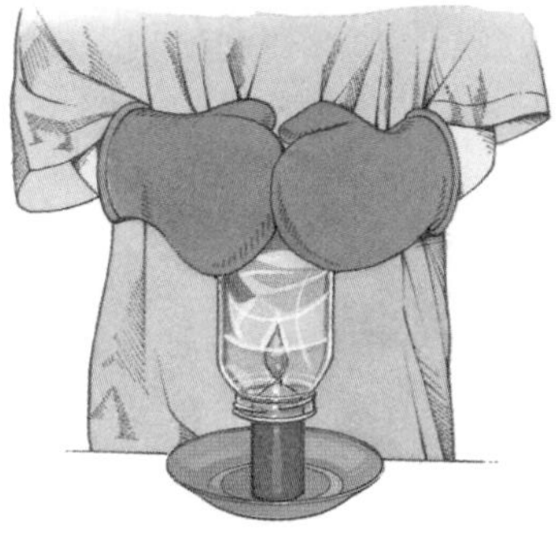

Figure 1

Figure 2

Now put the jar upside down in a bowl of water with the end of a plastic tube under its rim. Blow down the tube until the jar is full of exhaled air (Figure 2).

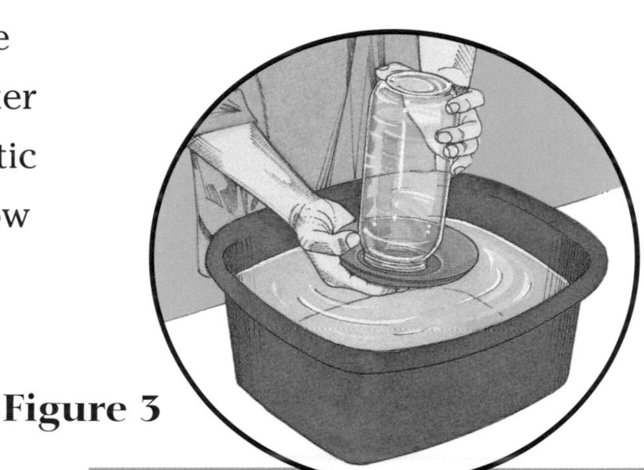

Figure 3

Quickly lift out the jar and cover the opening with a saucer (Figure 3). Light another candle and, wearing oven gloves, put the jar over it (Figure 4). Time how long it takes for the candle to go out. Does the candle burn longer in the first jar of air than in the jar of exhaled air?

Figure 4

WHAT THIS SHOWS

There is more oxygen in air you inhale (breathe in) than air you exhale (breathe out). Inhaled oxygen goes into your blood and all around your body. It is used up by your body and carbon dioxide waste is exhaled.

BREATHING RATE

Find a stopwatch or a watch with a second hand. Count how many breaths you take in one minute sitting still, walking, and then running. Your breathing gets faster the harder you work.

WHAT THIS SHOWS

The more energy you use, the more oxygen your body needs. You breathe more quickly when you are running than when you are sitting because you need to take in extra oxygen.

Take a deep breath!

You can probably only swim a few strokes under water before you have to come up for air. Turtles can hold their breath and stay under water for several hours.

MISTY MIRROR

Your breath is warm and full of droplets of water you can't usually see. Breathe on a cold mirror, and the droplets in your breath will cover the mirror in mist. When you breathe on a warm mirror, the droplets will not turn into a mist, but they are still in the air.

All animals need oxygen to survive. They get the oxygen they need from the air or from water. Lungs and gills are equally important for getting rid of waste gas.

Feeding

All animals need food and water to give them energy to live and grow. Digestion is the way your body breaks down food and uses the goodness from it. Your body then gets rid of any waste. For a healthy diet humans need carbohydrates, such as bread or pasta, for energy. Fruit and vegetables provide vitamins and minerals. Protein, such as eggs and meat, is needed for growth and healing, and a little fat is needed for warmth.

Digestive juices at work

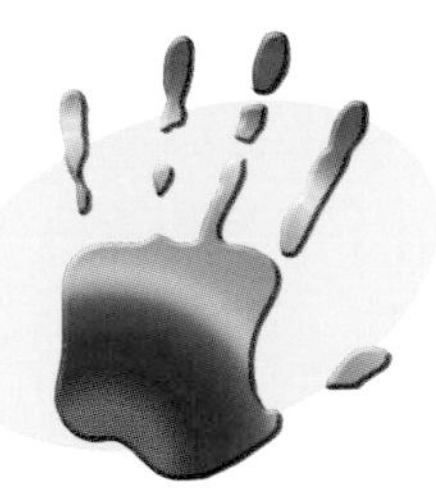

METHOD NOTES
The bread will taste sweet more quickly if you chew a small piece.

Materials
- a potato
- a slice of bread
- iodine
- a knife
- a dropper

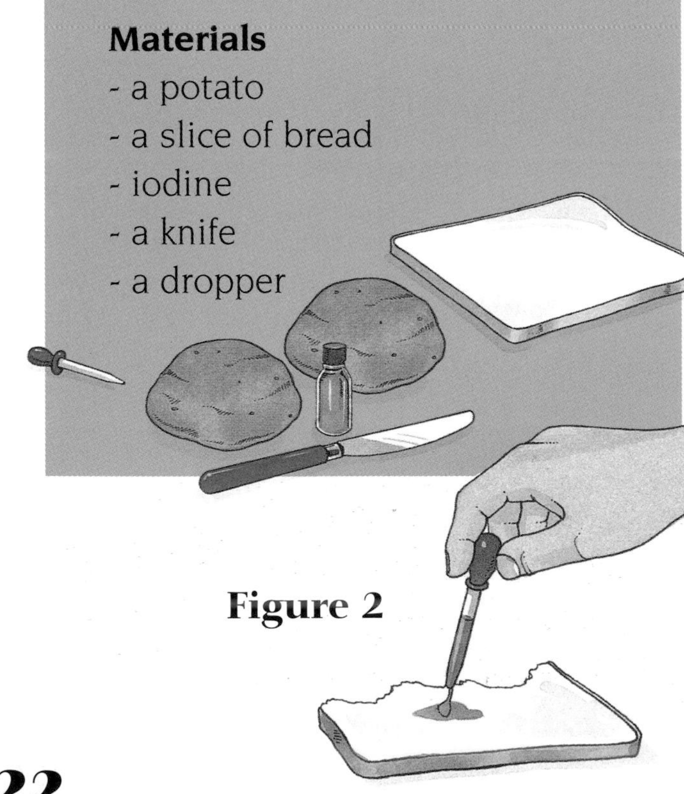

Figure 2

1. Tear a piece of bread in half (Figure 1). Test one half of the bread for starch by dropping a few drops of iodine onto it (Figure 2). The starch should turn the iodine blue immediately.
2. Chew the other half of the bread for a couple of minutes (Figure 3)—the half without iodine on it! Keep chewing until the bread tastes sweet.

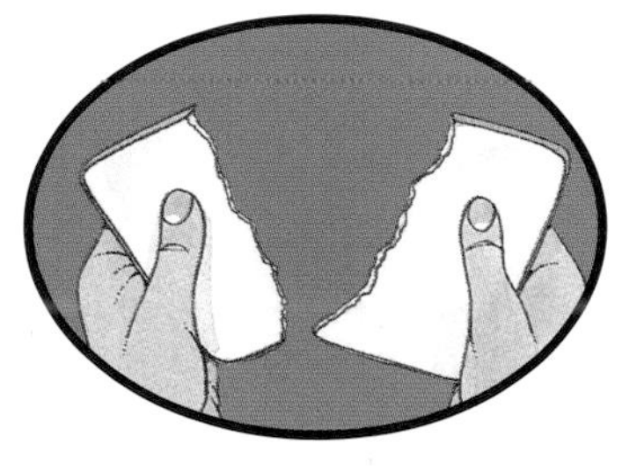

Figure 1

Figure 3

3. For the next experiment, cut a potato in half, spit some saliva onto one half, and smear it all over the cut surface (Figure 4).

4. Leave the saliva to do its work. Now put a drop of iodine on both halves.

5. Compare the color of the iodine on the two halves. The iodine on the piece of potato with saliva on it remains a brown color. The iodine on the other piece should turn blue (Figure 5).

Figure 4

Figure 5

WHAT THIS SHOWS

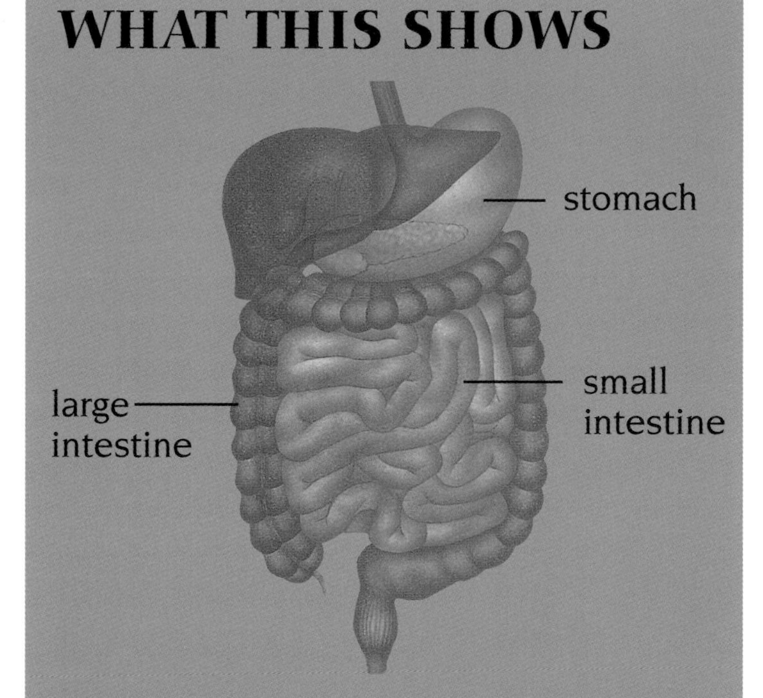

Enzymes, called amylases, in your saliva start to break down the starch in food and turn it into sweet-tasting sugars. Your food travels down a tube into your stomach where more enzymes work on it. As it travels through your small and large intestines, nutrients pass into your body, and you eventually get rid of any waste.

The incredible journey

Food takes up to 36 hours to go all the way through your digestive system. It stays in your stomach for about 3 hours and then travels along 30 feet (9 m) of tubes.

Feeding

Herbivores are animals that eat only plants, carnivores eat only meat, and omnivores eat both plants and meat. Humans are omnivores, but some people choose to eat just plants.

MAKE A TONGUE MAP

Your tongue detects only four basic tastes—sweet, sour, salty, and bitter. Being able to taste your food helps to save you from eating things that are harmful. Spoiled food often tastes so bitter that you want to spit it out at once.

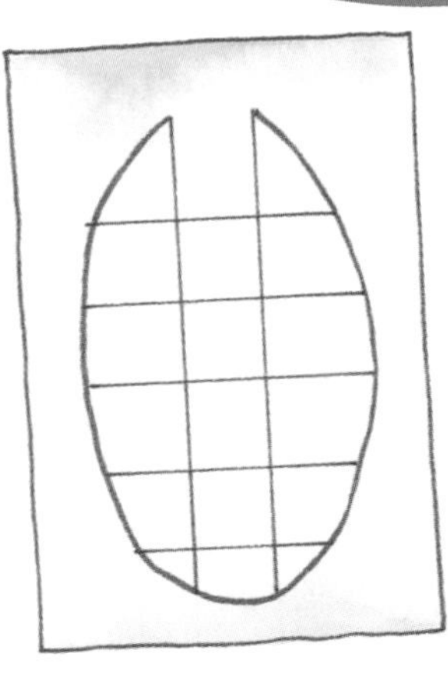

Figure 2

1. Pour some water into 3 glasses. Dissolve a little sugar in one, salt in the second, and coffee in the third. In a fourth glass put lemon juice (Figure 1).
2. Draw a map of a tongue like the one in Figure 2.
3. Blindfold a friend and drop a little of each mixture onto different parts of his or her tongue (Figure 3).
4. Mark on the tongue map where on the tongue your friend tastes each mixture most strongly (Figure 4).

Figure 1

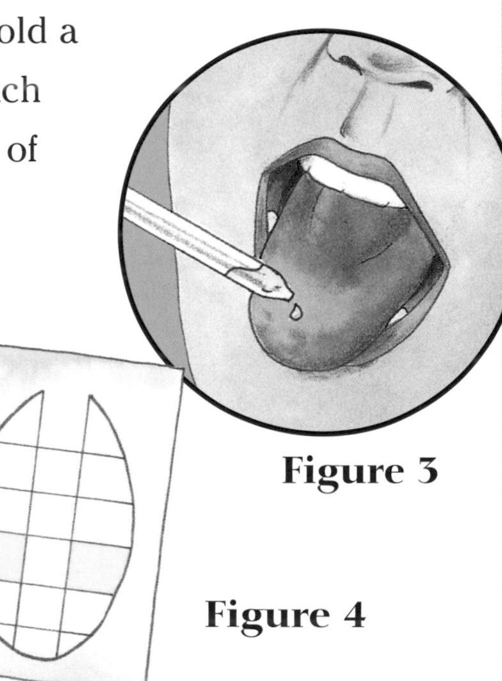

Figure 3

Figure 4

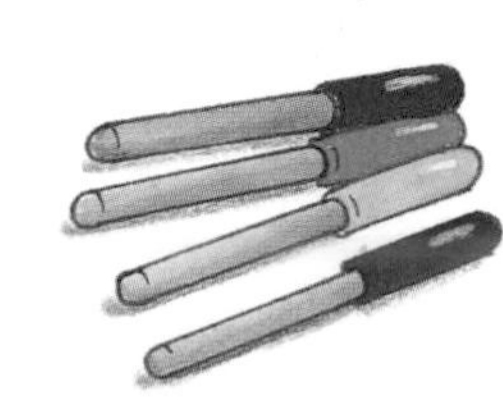

WHAT THIS SHOWS

Different parts of your tongue detect different tastes.

Stick out your tongue and look in the mirror. It is covered with small bumps called papillae, or taste buds. They detect tastes and send this information to your brain.

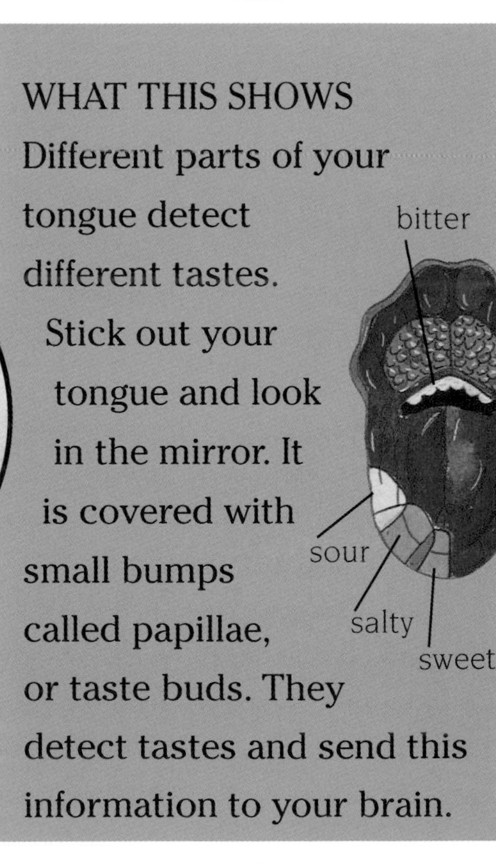

THE AMAZING GURGLING STOMACH

Listen to the noises your stomach makes

Use a tape recorder and a microphone to record the rumbles and gurgles you sometimes hear coming from somewhere inside you. Turn up the volume and replay the sounds to amaze your friends.

WHY IT WORKS

The noises are made by gas bubbling through your intestines as you digest your food. Broccoli and beans create very interesting sounds!

WHAT THIS SHOWS

Muscles squeeze and push your food on its long journey through the tubes of your digestive system. This muscle movement is called peristalsis. It is so strong that food would go into your stomach even if you ate standing on your head.

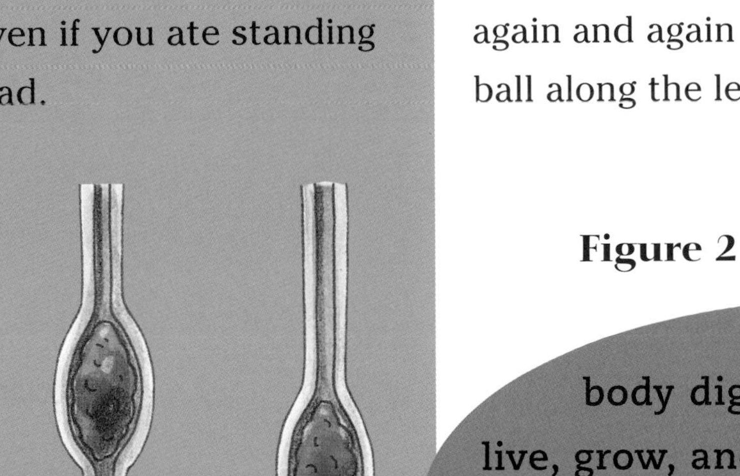

STOMACH MOVEMENTS

Push a small ball into a leg of an old pair of panty hose (Figure 1). Make a circle with your finger and thumb and squeeze and release them again and again to push the ball along the leg (Figure 2).

Figure 1

Figure 2

Your body digests food for you to live, grow, and keep healthy. Food that isn't needed by the body, or that is harmful, is gotten rid of at the end of the digestive system.

Movement

Every kind of creature in the animal kingdom moves. Some crawl, some fly, some swim, and some walk on legs or use their arms to swing through trees. Movement goes on inside bodies, too. The heart beats, blood pumps through veins, and lungs breathe air in and out. You have a skeleton of strong bones to support you and to protect your brain, heart, and lungs. Your skeleton bends at places called joints, where your bones join together. Muscles pull your bones to make them move.

Explore how joints move

METHOD NOTES
See which ways all the joints in your body move before you make these model joints.

Materials
- 2 empty toothpaste boxes
- an apple
- a pencil
- adhesive tape
- a cup

1. Put two rectangular boxes end to end and tape them together across one side only (Figure 1). Bend them at the place where they are joined, and notice how they will bend only in one direction (Figure 2), like your knee joint (Figure 3).

Figure 1

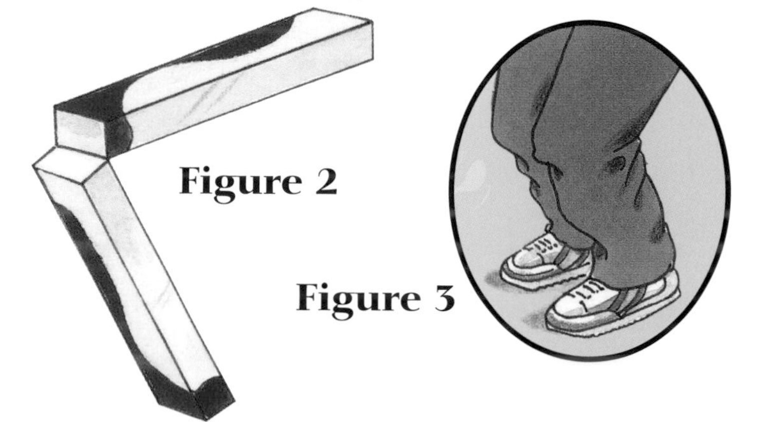

Figure 2

Figure 3

Figure 4

2. Push the pointed end of a pencil firmly into an apple to show your upper arm bone (Figure 4).

3. Put the apple into a cup to show how your upper arm bone fits into your shoulder bone. Move the pencil around, keeping the apple in the cup (Figure 5). It should swivel like your shoulder joint.

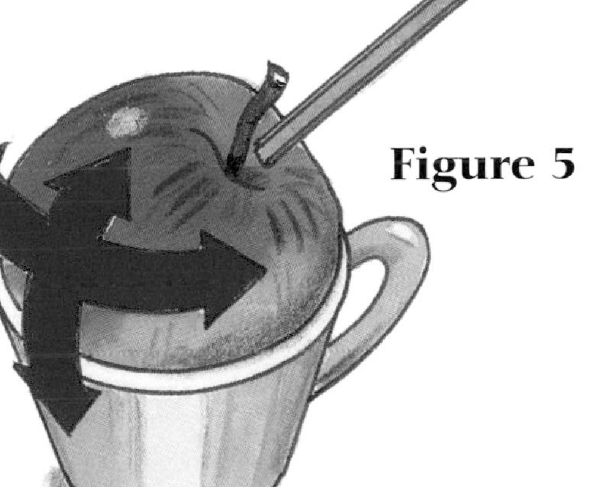

Figure 5

WHAT THIS SHOWS

You have hinge joints in your knees (1) and elbows, and ball and socket joints in your shoulders (2) and hips. You have sliding joints in your ankles and toes, pivot joints in your wrists, and saddle joints in your thumbs.

Each kind of joint moves in a different way. Your knee joints work like door hinges. They only bend in one direction and lock when you stand up straight. Your shoulders are ball and socket joints that let you circle your arms.

Making faces

More than 40 muscles in your face move to show exactly how you feel. But you can frown, smile, wink, and look surprised without moving a single joint.

Movement

You could not move without muscles to pull your bones. Muscles are attached to your bones by straps called tendons. Your bones are light and very strong for their weight.

HOW ARE BONES STRONG AND LIGHT?

Make a cylinder by taping together the ends of a rectangular piece of paper (Figure 1). Test the strength of the cylinder by putting books on top of it one by one (Figure 2). How many books can it support before it collapses?

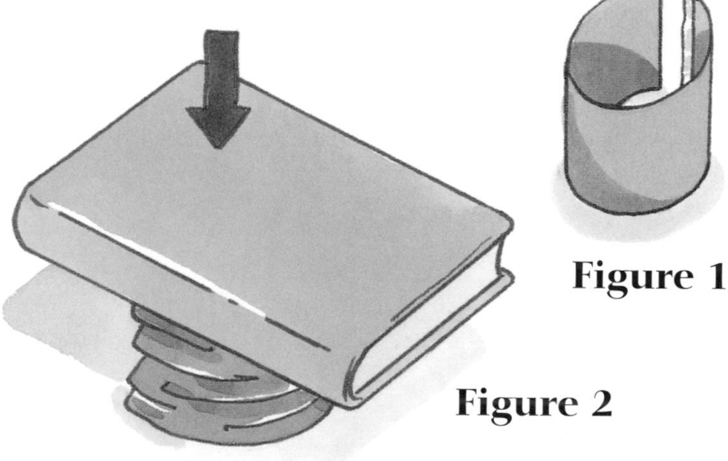

Figure 1

Figure 2

Make another paper cylinder in the same way. Make small paper cylinders the same height as the big cylinder. Pack the big cylinder with the small ones (Figure 3). Feel how light it is. Now test its strength by seeing how many books it can support (Figure 4). It should be much stronger than the empty cylinder. You have made a surprisingly strong structure from pieces of paper.

Figure 3

Figure 4

WHAT THIS SHOWS

The outer layer of bone is called compact bone. It is made of hundreds of cylinders. The inside layer of bone is called cancellous bone and is full of tiny hollow tubes. These two layers make bone light and very strong, just as the cylinders you have made make your "bone" strong and light.

cylinders

TEST YOUR MUSCLE STRENGTH

Some of your muscles need to be stronger than others. Squeeze a bathroom scale as hard as you can between your knees and then your hands. Make a note of the reading on the scale. Which muscles are more powerful?

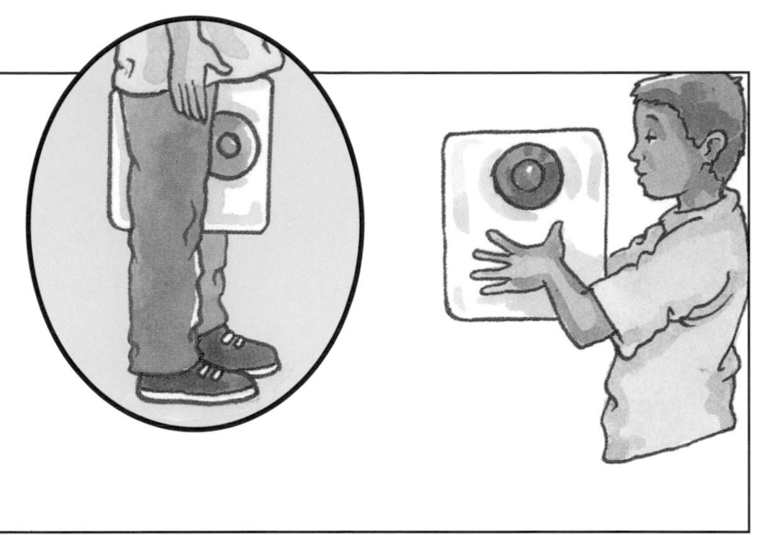

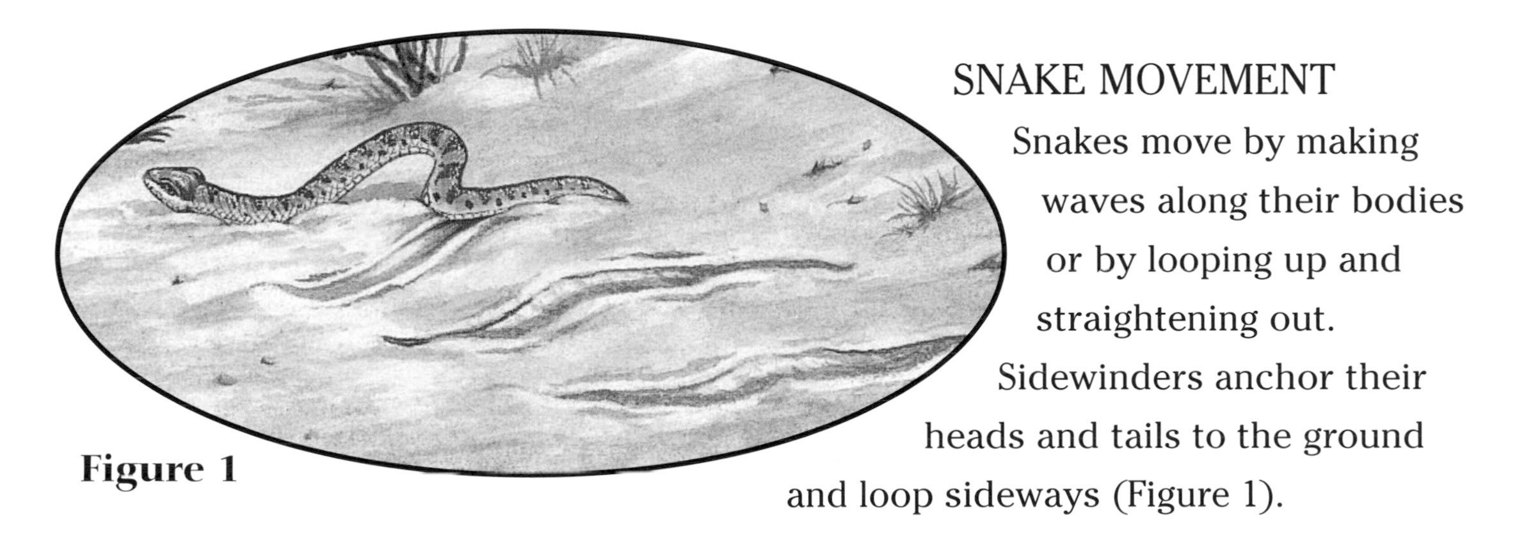

Figure 1

SNAKE MOVEMENT

Snakes move by making waves along their bodies or by looping up and straightening out. Sidewinders anchor their heads and tails to the ground and loop sideways (Figure 1).

You can make your own “snake” to try out these movements. Crumple newspaper to make about ten balls. Fill the leg of a pair of panty hose with a row of the newspaper balls. Make waving, looping, and sidewinding movements. Try them on a shiny floor and a carpet to see how a rough surface helps a snake to move forward.

Movement goes on all the time inside and outside your body. Your bones bend at the places where they join called joints, and muscles pull your bones to move you.

Reproduction

The Earth is teeming with life. Every kind of creature can reproduce—make a new life like itself—so that life on Earth can go on. The process of birth, growing up, reproducing, and dying is called the life cycle. A new life begins with an egg inside a female's body. It starts to grow when it is joined by a seed from a male.

Make a birdhouse to attract nesting birds

Materials

- a piece of wood $51\frac{1}{2}$ in by 6 in and about $\frac{3}{4}$ in thick.
- a saw and a ruler
- strong waterproof glue
- a marker and a brush
- a hammer and nails
- wood preservative
- a wooden stick

METHOD NOTES

Don't worry about gaps in the joints. They make good ventilation.

Ask an adult to help you with this project.

1. Ask an adult to cut the wood into six pieces according to the measurements shown below (Figure 1). The sides are pieces 1 and 2, the back is piece 3, the front is piece 4, the base is piece 5, and the lid is piece 6.

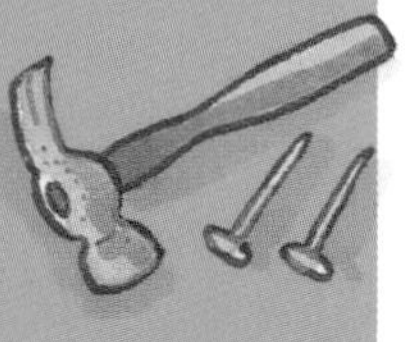

Figure 1

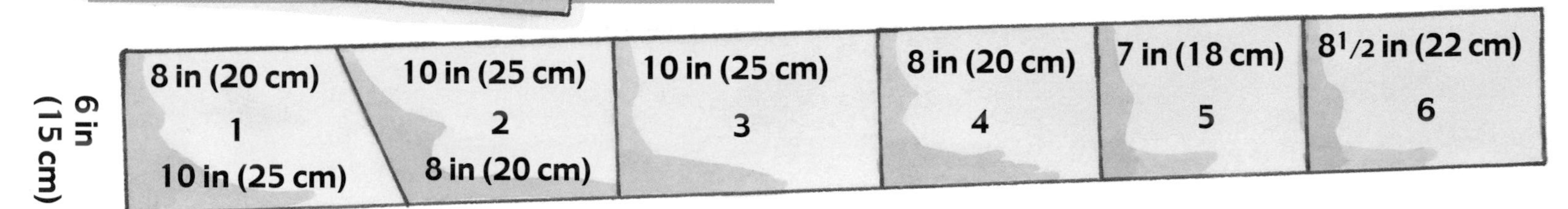

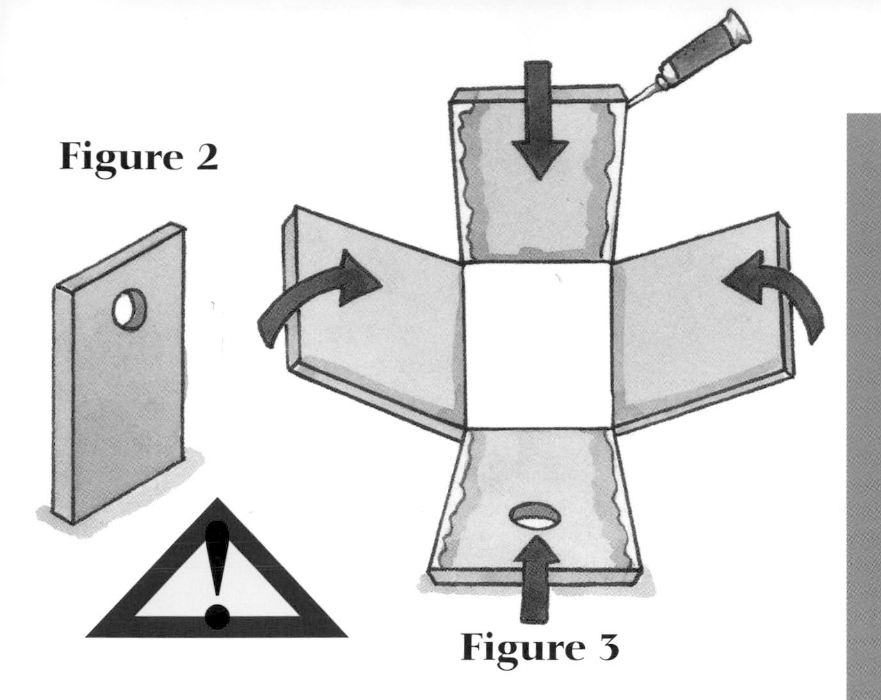

Figure 2

Figure 3

2. Ask an adult to cut a hole $1\frac{1}{4}$ in (30 cm) across in the front piece above the center (Figure 2).
3. Glue the pieces onto the base, as in Figure 3. Let the sides and base dry before you glue on the roof (Figure 4).
4. Paint the outside with wood preservative to protect the birdhouse from the weather.

WHAT THIS SHOWS

Birds need a safe place to make a nest and lay their eggs in spring. It must be off the ground, away from predators. All birds start life in an egg laid by the mother bird. The adult bird sits on the egg to keep it warm. The egg hatches into a hungry nestling that is fed by the parents. The young bird becomes a fledgling. It learns to fly and look for its own food. It eventually becomes an adult and finds a mate. The female lays eggs in the spring.

Figure 4

5. Carefully nail a stick of wood to the back of the birdhouse and then nail the stick to a tree trunk (Figure 5).
6. Put moss, bark, leaves, and twigs on the ground under the house for birds to use as nesting material (Figure 6). Watch for birds to nest in the house in the spring.

Figure 5

Figure 6

Reproduction

Most baby mammals grow inside their mothers and are then born alive—not in an egg. Reptiles, birds, fish, and insects lay eggs. The baby develops inside the egg until it hatches.

EXAMINE THE INSIDE OF A HEN'S EGG

Crack open a hen's egg onto a white plate (Figure 1). Examine it with a magnifying glass (Figure 2). The egg shell protects the growing baby bird. The yellow yolk is food for the growing chick. The egg white acts as a cushion to protect the embryo (baby bird). In eggs that have been fertilized by a rooster, there will be a red spot in the yolk. This is the embryo. There will not be an embryo in your egg.

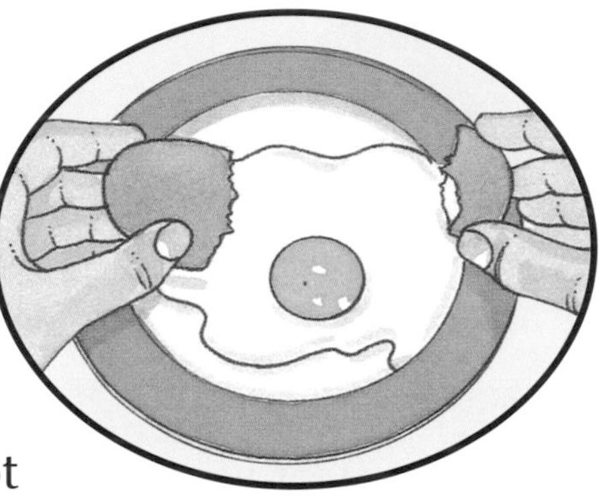

Figure 1

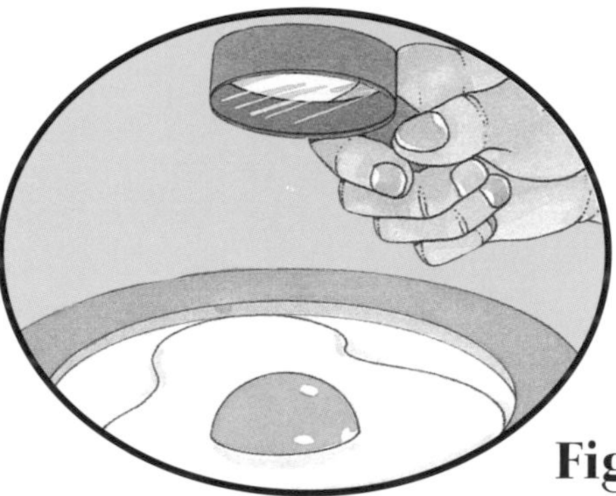

Figure 2

THE AMAZING ROLLING TONGUE

See how things are passed to you from your parents

Find out which of your friends can spread their fingers 2 by 2 into a V shape and which can roll their tongue. Some will do it easily and some won't be able to, however hard they try.

WHY IT WORKS

Rolling your tongue or spreading your fingers are abilities you inherit from one or both of your parents, just as you may inherit the color of your eyes or the color of your hair.

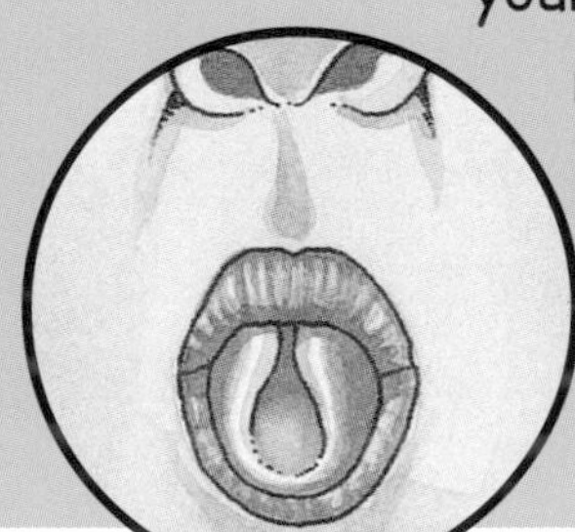

FINGERPRINTS

1. Draw separate boxes on a piece of paper. Press your thumb onto an ink pad (Figure 1).
2. Print it by pressing it onto paper in one of the boxes and gently rocking it from side to side (Figure 2). Write your name above it.

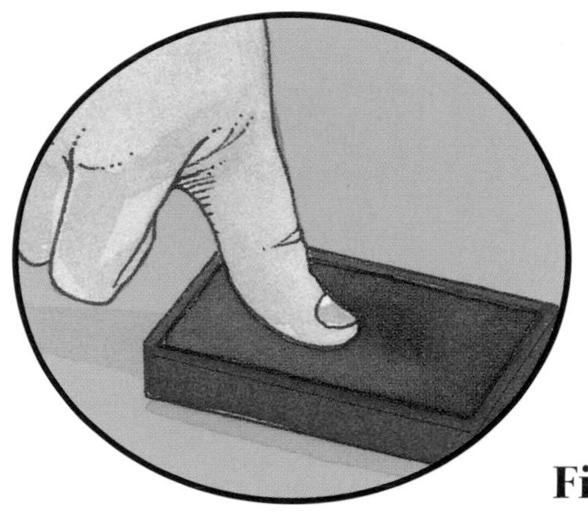

Figure 1

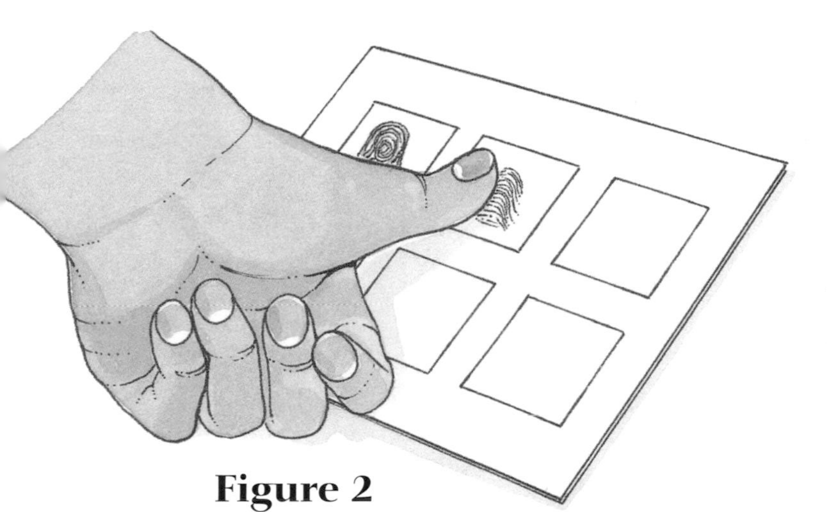

Figure 2

3. Make a collection of your friends' thumbprints and examine them through a magnifying glass.
4. Try to identify the patterns and write the name of the pattern under each print.

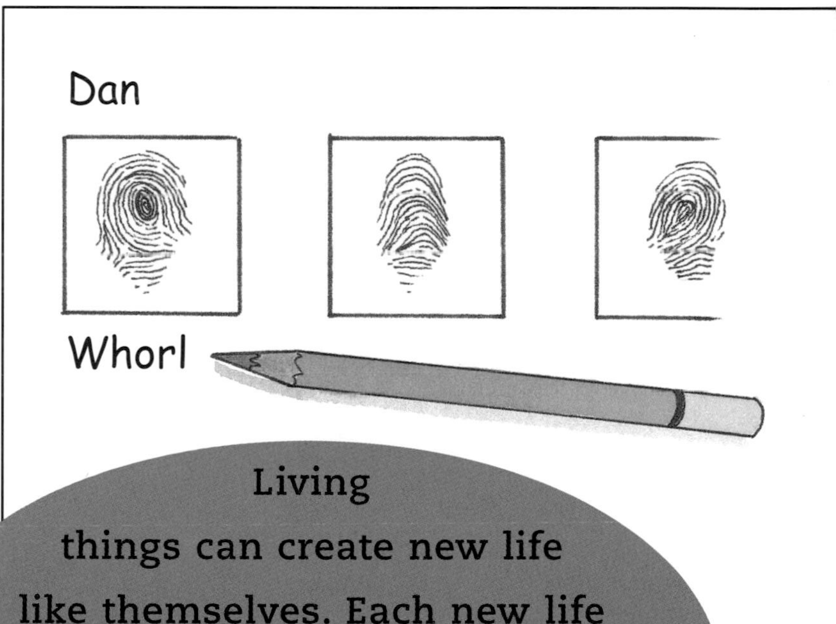

WHAT THIS SHOWS

Police search for fingerprints to identify a criminal because everyone has his or her own unique set of prints. Although everybody's prints are different, they are all made up of just four patterns: loop (1), arch (2), composite (3), and whorl (4).

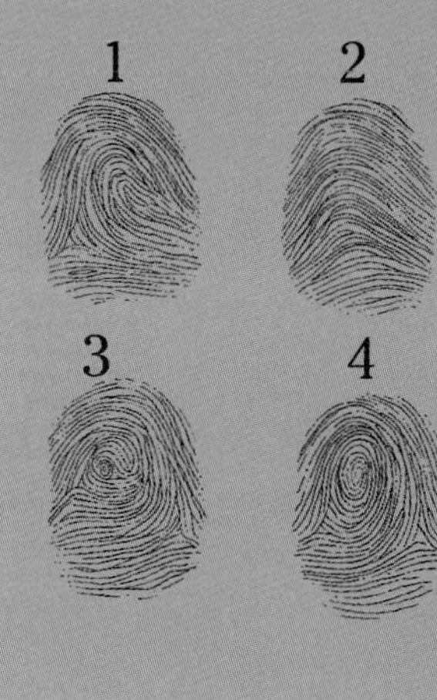

Living things can create new life like themselves. Each new life begins a life cycle. You inherit some of your features from your parents, but there is no one exactly like you.

Growth

When you become an adult, you will stop growing taller, but things like your fingernails and hair never stop growing. A hairless baby kitten, a featherless baby bird, and a helpless baby human being are not very different from the adults they grow into. A baby whale looks just like a smaller version of its mother. Some creatures undergo a much more dramatic change called metamorphosis. For example, a butterfly spends the first part of its life as a caterpillar, and a frog starts life as a tadpole.

Compare head size with body size

METHOD NOTES
Try to do this with a toddler, someone your age, and an adult.

Materials
- a roll of kraft paper
- a marker
- scissors
- sticky tape
- a ruler

1. Measure your friend's head from chin to crown (Figure 1). Now measure his height.
2. Spread some kraft paper on the floor. Tape down the corners. Ask your friend to lie on it and then draw around him (Figure 2).

Figure 1

Figure 2

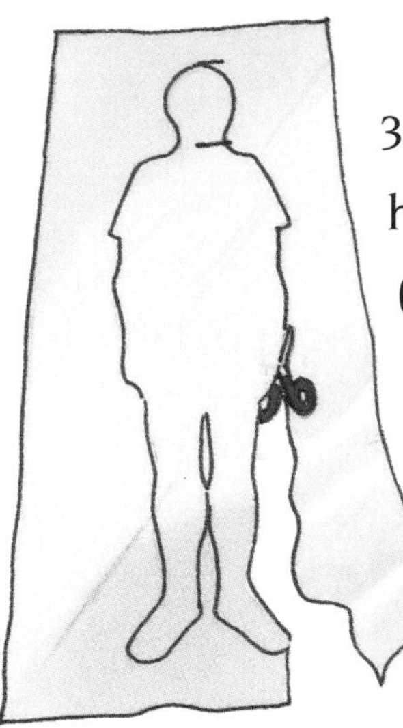

Figure 3

3. Mark on the length of the head and cut out the shape (Figure 3). Fold down its head at the neck. Keep folding the paper figure using the head as a guide for the size of the rest of the folds. Also do this for a toddler and an adult.

4. Unfold the figures. They should fold into about 4 sections for a toddler (Figure 4), 5 or 6 for your friend (Figure 5), and about 8 for an adult.

WHAT THIS SHOWS

A toddler's head is nearly as large as a grown-up's head, but its body is much smaller. Humans are born with a well-developed brain. The toddler's head makes up about one quarter of its whole body length because the body still has a lot of growing to do! By the time the toddler is an adult, its head will be only about one eighth of its total body length.

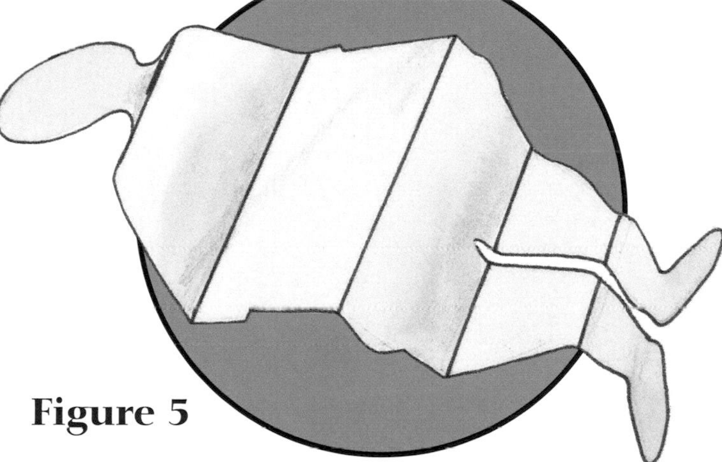

Figure 5

Figure 4

Quick, run away!

A baby deer can stand up and run around almost immediately after birth. This means it can run away from predators. Human babies learn to walk many months after birth.

Growth

Baby mammals drink their mother's milk. It is full of the nutrients they need to grow. Milk provides calcium, for healthy teeth and bones, protein, fats, and other minerals and vitamins.

MAKE BUTTER FROM CREAM

Cream, butter, cheese, and yogurt are all made from milk. You can make butter at home. Pour a little cream into a small, clear plastic bottle and screw the lid on tightly. Shake the cream in the bottle (Figure 1) until it starts to separate into thin liquid and small lumps of butter (Figure 2). The fat in butter makes a greasy mark on clean paper (Figure 3). The fat in milk is important to the growth of young mammals. The milk of cold-weather mammals is mostly fat, and helps their young grow layers of fat to protect them from the cold.

Figure 1

Figure 2

Figure 3

PHOTOGRAPH FUN

We all change as we grow older. Collect photographs of adults you know taken when they were children. Stick them on cardboard. Write the names of the adults in a different order down the side. Give a prize to the person who matches the most names to the right pictures.

TADPOLE VIEWER

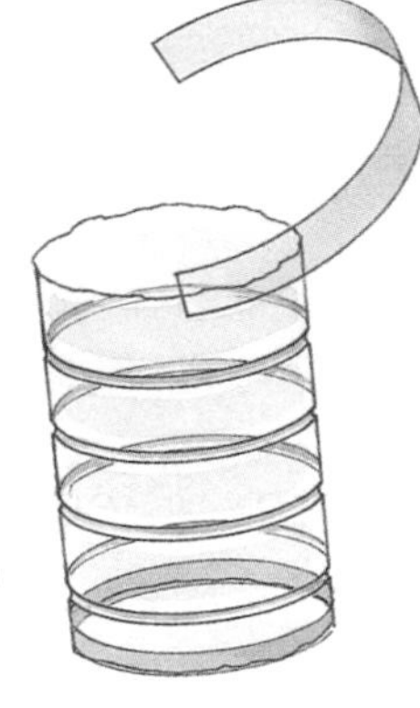

Figure 1

1. Find a strong, clear plastic bottle. Ask an adult to cut off the top and bottom ends, and tape over the rough edges (Figure 1).
2. Pull plastic wrap tightly over one end of the bottle and attach it with a rubber band to keep it from slipping off in the water (Figure 2).

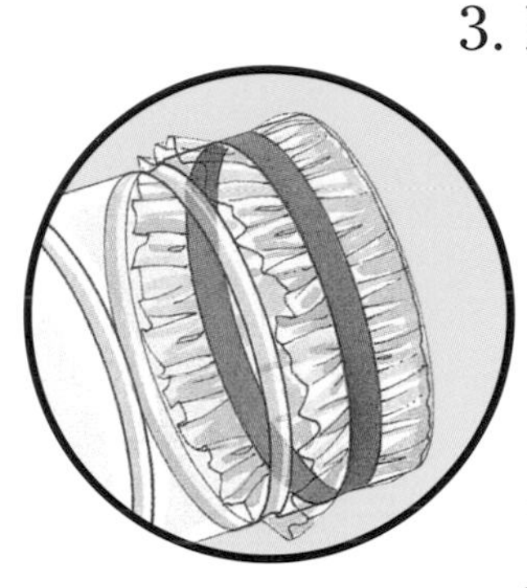

Figure 2

3. Put the end covered in plastic wrap in a pond. You should be able to see what's going on under water (Figure 3). Use your viewer to watch tadpoles in the spring. **Be very careful near water.**

Figure 3

WHAT THIS SHOWS

The tadpole stage is just one of the stages in the life cycle of a frog. Frogs lay their eggs in the water. The eggs hatch into tadpoles, which live under water. The tadpoles lose their tails, grow legs, and become frogs, which live on land and in the water.

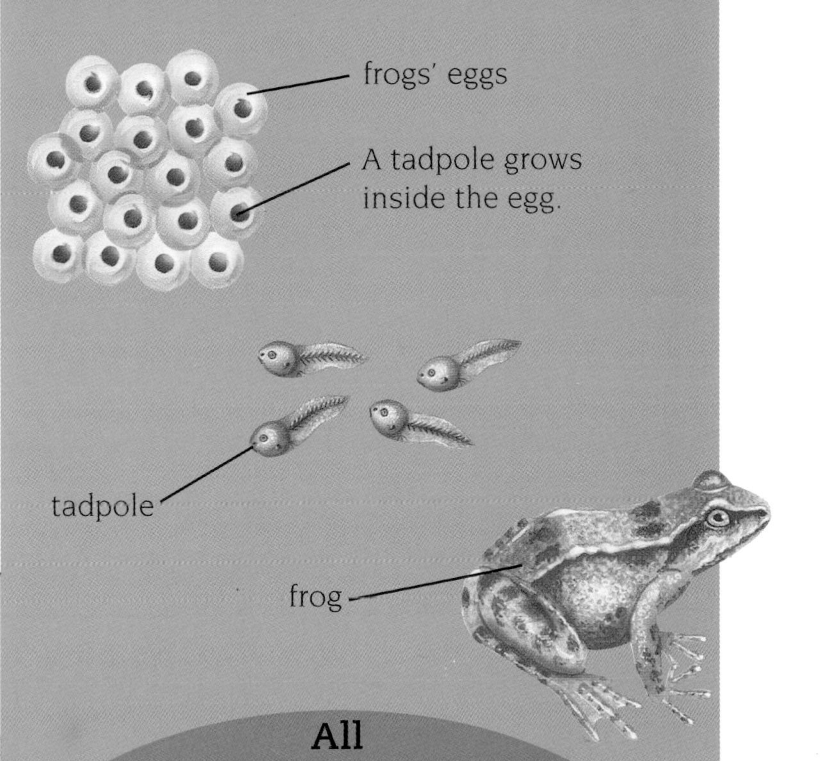

All animals grow. Some are born looking like their parents, others look very different. Animals develop at different rates. Humans take years to reach adulthood, yet some animals take only a few months.

Senses

You use your five senses—sight, hearing, smell, taste, and touch—to make sense of what is going on in the world around you. Your eyes, ears, nose, mouth, and skin are your sense organs. They pick up information and send messages to your brain. Many animals have highly developed senses. A dog's sense of smell is estimated to be one thousand times stronger than yours, and cats have excellent sight, even at night.

Find out how your eyes see

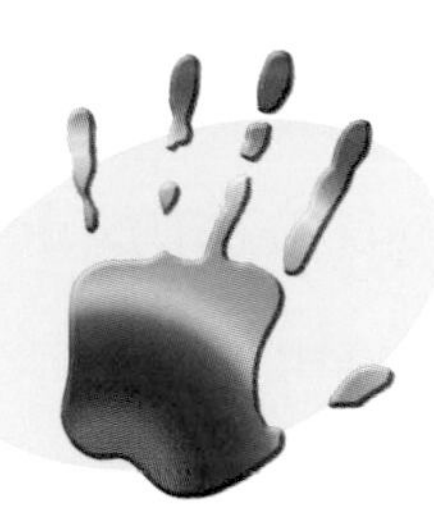

METHOD NOTES
If you use stiff cardboard, ask an adult to help you cut it with a craft knife.

Materials
- 1 sheet of cardboard
- 1 sheet of tracing paper
- a pencil
- a ruler
- glue
- scissors
- a magnifying glass

1. Draw a rectangle on the cardboard about $1^{1}/4$ in (3 cm) from the edges to make a frame.
2. Cut out the rectangle by making a hole in it with your scissors and then cutting along the lines (Figure 1).

Figure 1

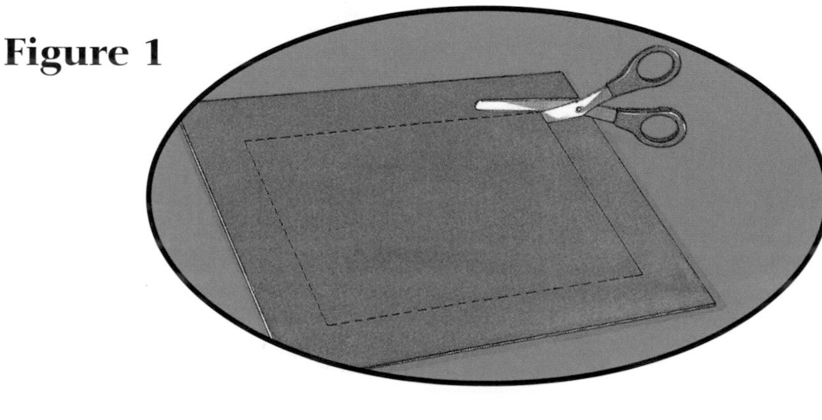

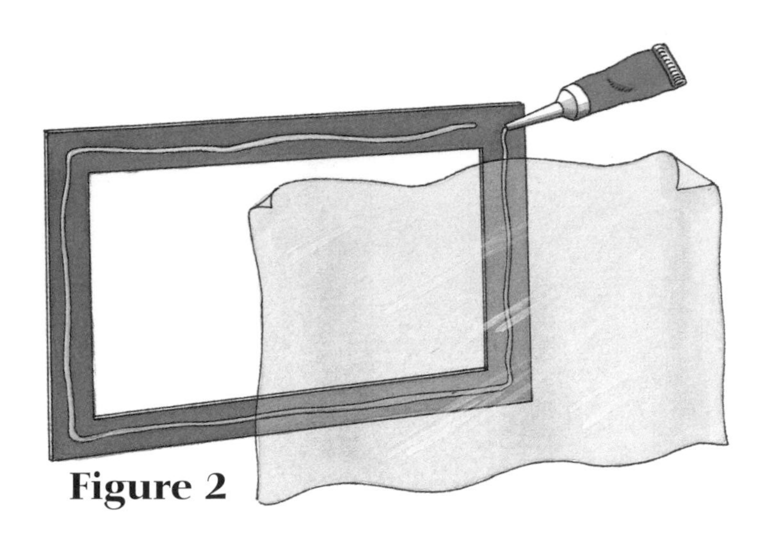

Figure 2

3. Put glue on one side of the cardboard to stick the tracing paper over the frame to make a screen (Figure 2). Make it as smooth and tight as you can.
4. Hold the frame up to a window or any bright object. Hold the magnifying glass beyond the frame and adjust it until you can see a clear upsidedown image of the bright object on the screen (Figure 3).

Figure 3

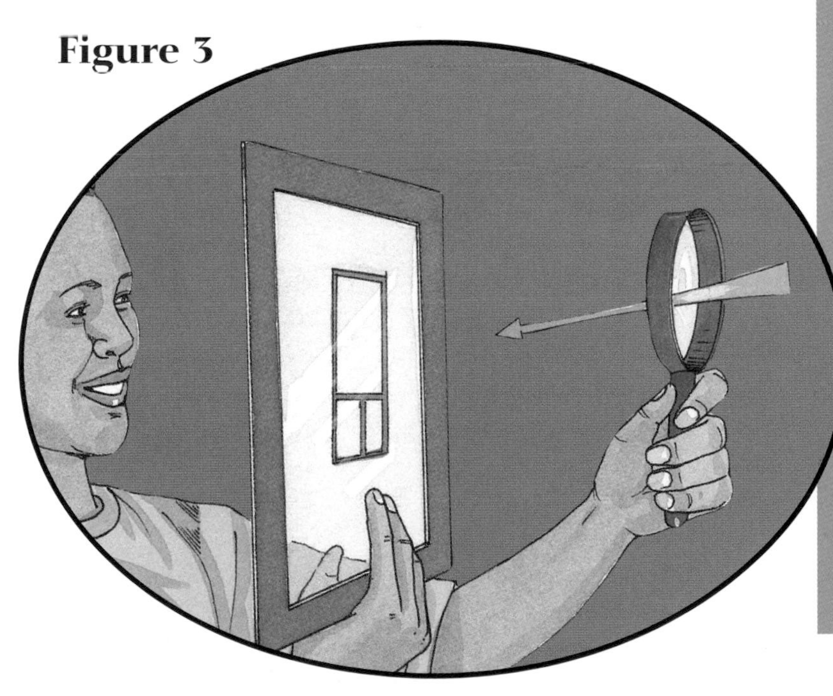

WHY IT WORKS

In this experiment, the magnifying glass represents the lens of your eye. The screen represents your retina at the back of your eye. Light rays bounce off the objects you look at and go into your eye.

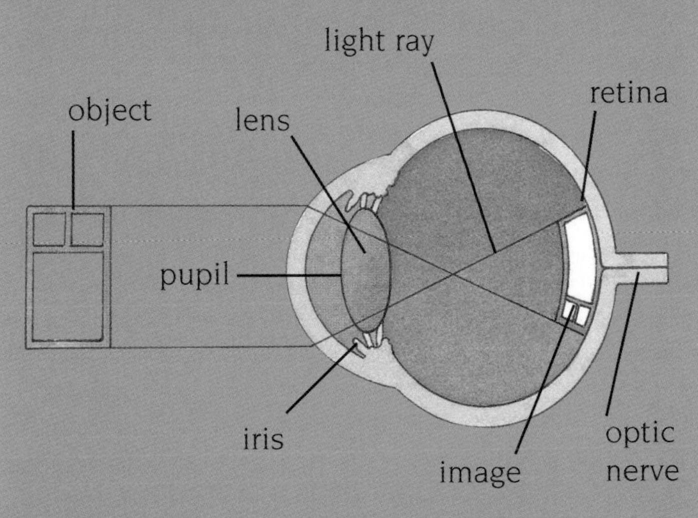

The iris (colored part) of your eye changes size to let more or less light through the pupil and into the eye. The lens focuses the light, making an upsidedown image on the retina at the back of your eye. Your optic nerve picks up the information and sends it to your brain where it is sorted out into the image that you see.

Senses

If one of your senses is lost, sometimes you can use your other senses. You can use your sense of touch to read Braille and your eyes to see what people are saying (lip reading).

USE TOUCH, SMELL, AND TASTE
Put on a blindfold and ask someone to give you an unfamiliar object. Feel it, using your sense of touch, to get as much information about it as you can. Now draw it (Figure 1). Is your drawing anything like the real thing?

Figure 1

Still blindfolded, get a friend to feed you pieces of apple, potato, cheese, and chocolate while you hold your nose (Figure 2). Is it difficult to tell what you are eating?

Figure 2

WHY IT WORKS
You can taste food much better if you can also smell it because smell and taste are very closely linked. Your nose smells more flavors than your tongue can taste, so when you have a stuffy nose, many foods taste the same.

THE AMAZING MAGIC PEN CAP
See how two eyes are better than one

With one eye shut. hold up a pen in one hand and its cap in the other. Try to put the cap on the pen. It's difficult with one eye shut. Now open that eye and try again. Suddenly, it's easy!

WHY IT WORKS
You have two eyes to help you figure out distances. With one eye shut, you can't tell exactly where things are. Animals that are hunters have two eyes on the front of their heads so they can pinpoint their prey.

BLOW OUT A CANDLE WITH SOUND

Cut out 2 circles of rubber from a large balloon (Figure 1). Stretch them over the ends of a cardboard tube and secure them with rubber bands (Figure 2). Pierce one piece of rubber to make a little hole (Figure 3). **Ask an adult to help you light a candle.**

Figure 1

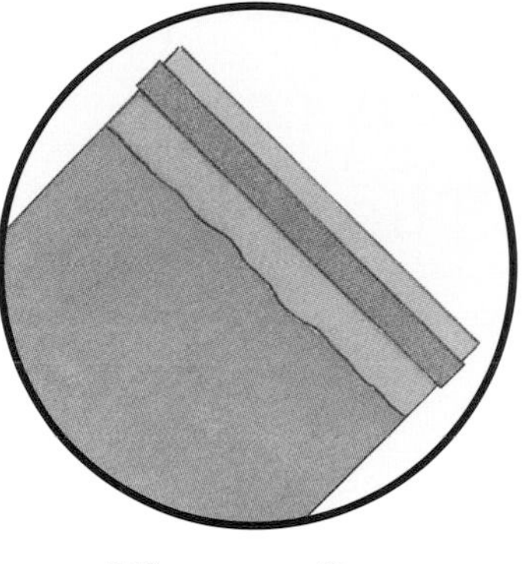

Figure 2

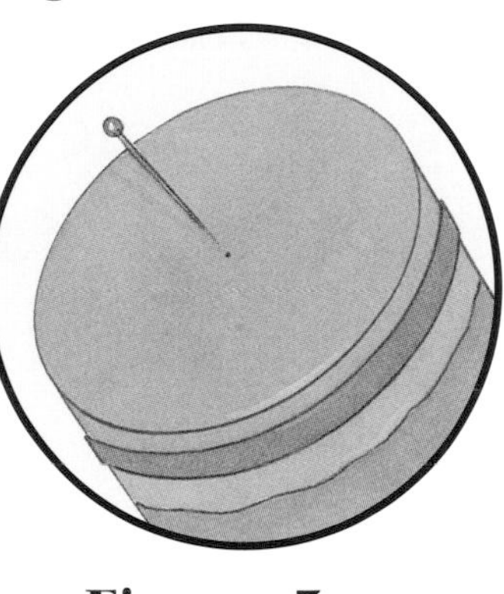

Figure 3

Hold the tube so that the hole in the rubber is in line with the candle flame. Hit the rubber at the other end of the tube (Figure 4) and watch the flame go out.

Figure 4

WHY IT WORKS

When something moves, it vibrates the air around it. The moving air travels as sound waves. The waves hit your eardrums, and you hear a sound. When you hit the rubber, it makes vibrations that you hear as a tapping sound. The vibrations move along the tube and push air out through the hole. This blows out the candle flame.

You see with your eyes, hear with your ears, smell with your nose, taste with your tongue, and touch with your skin to sense the world around you.

Adaptation

The environment is everything that surrounds us. Animals live in every kind of environment from the bottom of deep seas to high up in mountains, from icy polar regions to hot, dry deserts. In order to survive in their environments, animals often have to make changes. City foxes have changed their natural diet, and Arctic foxes change the color of their coats from brown to white when the snow falls. These changes to fit the environment are called adaptation.

Discover what the imprint of your teeth looks like

METHOD NOTES
Make sure you keep kneading your dough until it is no longer crumbly.

Materials
- 1 cup flour
- 1 cup water
- 1/2 cup salt
- 1 teaspoon cooking oil
- a mixing bowl
- a jug
- a wooden spoon

1. Measure the flour, water, salt, and oil into a mixing bowl. Mix all the ingredients together with a wooden spoon (Figure 1).
2. Dust your hands with flour and knead the mixture with your hands (Figure 2) until it is elastic and does not crumble.

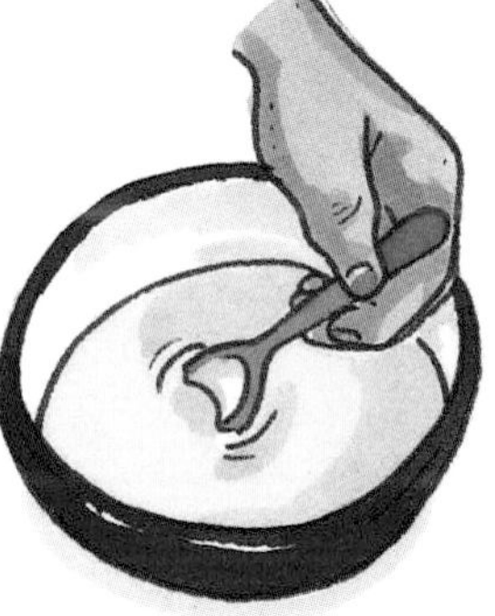

Figure 1

Figure 2

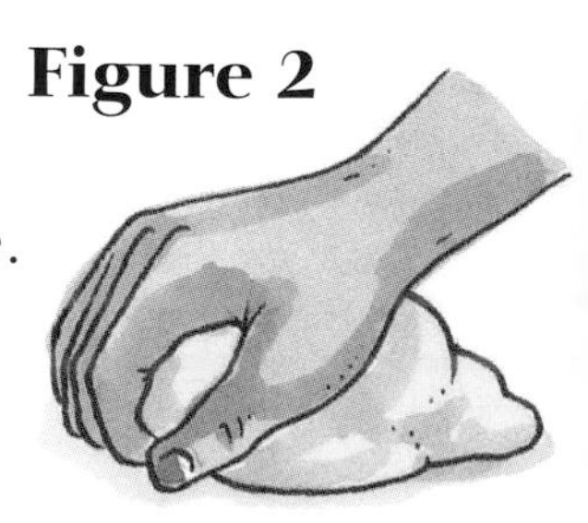

3. Break off some dough and roll it into a ball. Flatten the ball into the size and shape of a small cookie.
4. Moisten your teeth with your tongue so the dough doesn't stick to them. Now bite into the dough just hard enough to make an imprint of all your teeth, top and bottom.
5. Carefully remove the dough without spoiling your teeth marks (Figure 3). Put the dough in a low oven until it is dry.

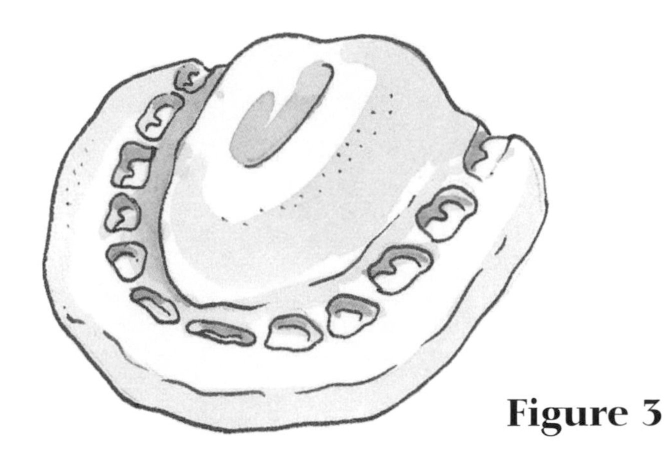

Figure 3

Sharks' teeth—vital for survival!
Sharks could not survive without their very sharp teeth. So, if one breaks, a new tooth growing behind it moves forward to take its place.

WHAT THIS SHOWS

You have 3 types of teeth—incisors, canines, and molars. Can you see the imprints of the different teeth in the dough?

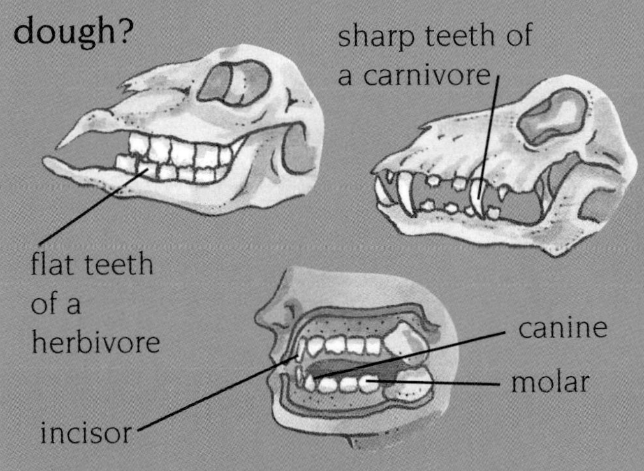

Incisors are at the front for biting, canines are for gripping and tearing, and molars at the back are for chewing. Animal teeth are adapted for the food they eat. Meat eaters have sharp teeth and herbivores have flat teeth.

Some animals are hunters and others are hunted. Hunters have keen sight and hearing, and a good sense of smell to track down their prey. Hunted animals are adapted for a quick escape.

MAKE AN EAR TRUMPET

Cut out a circle of thin cardboard the size of a dinner plate. Cut the circle in half (Figure 1). Roll one half into a cone and tape it together. Cut off one end so that it fits over your ear (Figure 2). Without using the ear trumpet, listen to a quiet sound such as a computer humming or a radio turned down very low.

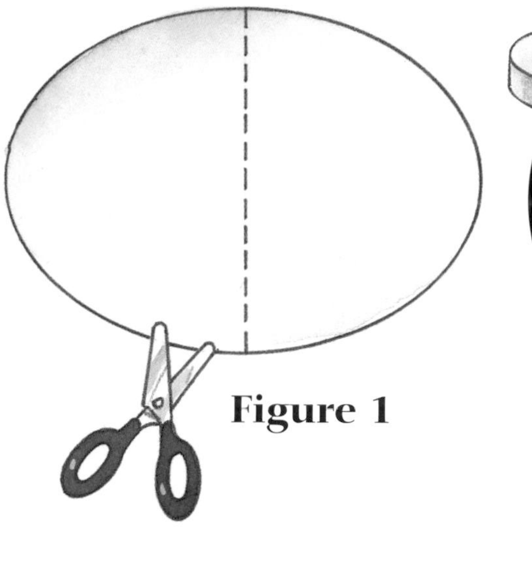

Figure 1

Figure 2

Figure 3

Now put the narrow end of your ear trumpet to your ear and point the wide end toward the sound (Figure 3). Listen carefully. Can you hear the sound more clearly through the ear trumpet? Move the trumpet around without moving your head to pick up other sounds.

WHY IT WORKS

Animals with very sharp hearing, like foxes, have big ears which they can prick up and swivel toward a sound. The part of your ear that you can see is called the earflap. It is designed for collecting sounds. The ear trumpet makes your earflap bigger. The bigger the earflap, the better you can hear.

COUNTERSHADING

Animals that live on grassy plains often have countershading—fur that is darker on top and lighter underneath. This helps them blend into the background, making it hard for hunters to see them. Make two animals from cardboard boxes and rolls (Figure 1). Color or paint one with a dark back and light belly and the other with a light back and dark belly (Figure 2). Put them both outside on a sunny day to find out which is more difficult to spot. Is the one with a dark top and a light belly harder to see?

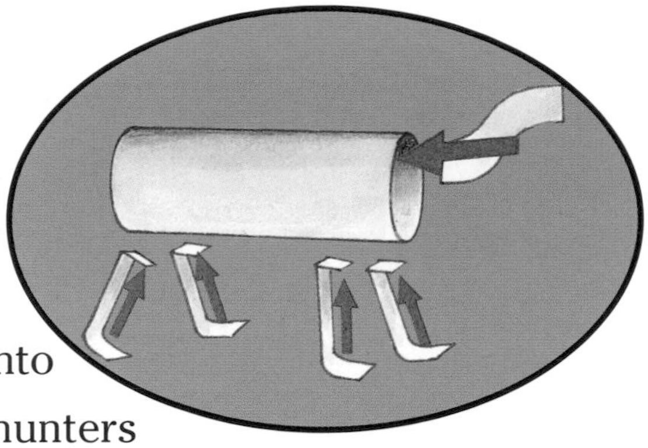

Figure 1

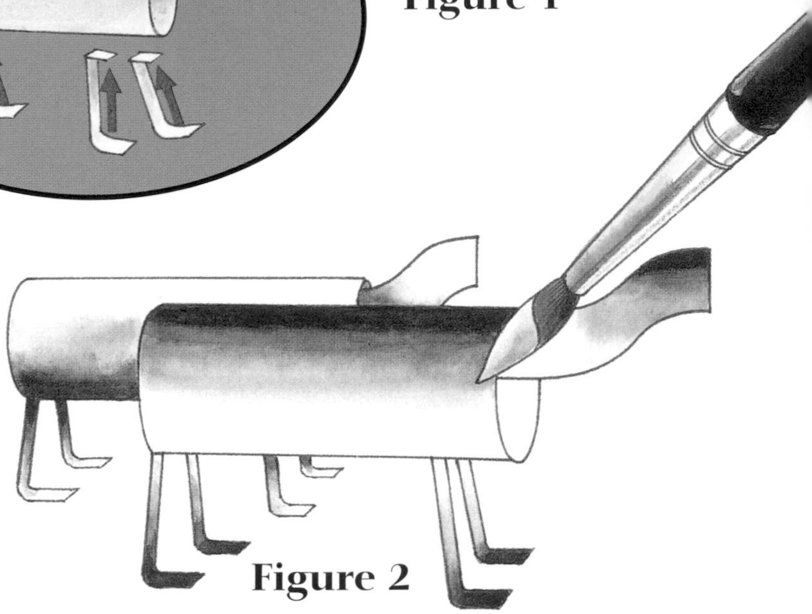

Figure 2

FINDING THE WAY

Birds migrate enormous distances to look for food, using the Earth's magnetic field to guide them. We use a compass, which always points to magnetic north, to help us find the way. Hide some treasure outside. Write instructions for finding it like those in Figure 1. Give your friends the instructions and a compass to help them find the treasure (Figure 2).

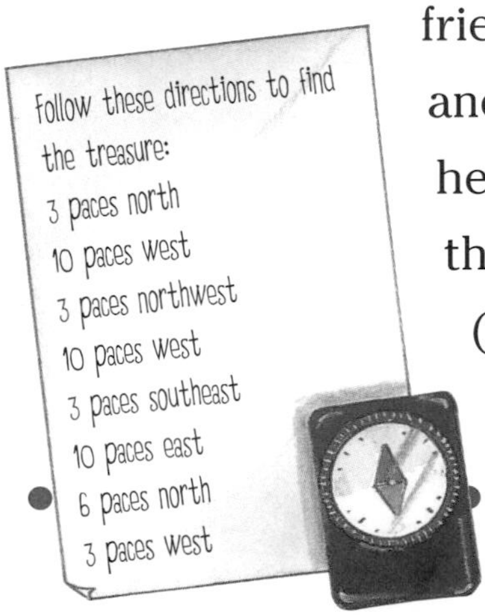

Follow these directions to find the treasure:
3 paces north
10 paces west
3 paces northwest
10 paces west
3 paces southeast
10 paces east
6 paces north
3 paces west

Figure 1

Figure 2

Animals adapt themselves to their environment in different ways. Adapting helps animals to live successfully so that life on Earth can continue.

Glossary

air
Air is a mixture of invisible gases all around us. Animals need to breathe air in and out of their lungs to live. They inhale (breathe in) a gas called oxygen and exhale (breathe out) a gas called carbon dioxide.

animal kingdom
Animals are one of four kingdoms of living things. Animals can be as different as a tiny insect and an enormous blue whale. Animals breathe, eat, get rid of waste, move, feel, grow, and make more life like themselves.

embryo
An embryo is an animal in the early stages of its development. Most mammal embryos develop inside their mother's body until they are ready to be born. Other animals lay eggs. The embryo develops inside the egg until it hatches.

environment
The environment is everything that surrounds us. A desert, a rainforest, and a city are all different kinds of environments. Animals have found ways of surviving in their environments by making changes to their diet, their coats, and the shapes of their bodies.

excretion
What happens when animals get rid of waste from their body. Waste is the part of food and drink that the body can't use.

exoskeleton
The hard covering around the body of arthropods that forms a protective armor around their soft, internal organs.

invertebrates
Invertebrates are animals without backbones. They are the largest group of animals in the animal kingdom and include worms, snails, and insects.

life cycle
All the stages a living thing goes through from birth to death. With each new life, another life cycle begins so that life on Earth can continue.

metamorphosis
The amazing change some creatures go through as they become adults. A caterpillar changes into a butterfly with wings. A tadpole living under water changes into a frog that hops about on land.

migration
The name for the long journey that birds, animals, and insects make to look for food.

predators
Animals that hunt and eat other animals. The hunted animals are called prey.

temperature
Temperature is a measure of how hot or cold something is. Human beings have a temperature of 98.6°F (37°C) and can keep their temperature constant whatever the weather. A reptile's temperature does not stay the same—it changes with the reptile's surroundings.

vertebrates
Vertebrates are animals with backbones. Vertebrates include animals of all sizes, such as whales, horses, and mice. Human beings are vertebrates, with 33 bones called vertebrae in their backbones.

Index

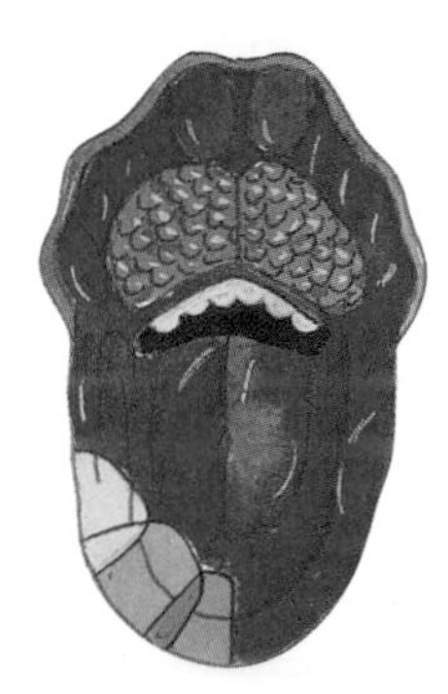